# RNA Sequencing:

## Principles and Data Analysis

Lara Ozkan

Copyright © 2020 Lara Ozkan

All rights reserved. No part of this book may be reproduced, stored in a retrieval system, or transmitted in any form or by any means, without the prior written permission of the published, except in the case of brief quotations embedded in critical articles or reviews.

ISBN: 9798680736388

RNA Sequencing: Principles and Data Analysis

# DEDICATION

I dedicate this book to my mother, father, Devran, and Eva. I want to thank my parents for expressing unwavering support for me and my countless endeavors.

## ACKNOWLEDGEMENTS

I'd like to thank my mentors Dr. Yagmur Kiraz, Dr. Yusuf Baran, Dr. Semir Beyaz, and Dr. Cem Meydan. Their vision and dedication is truly inspiring.

I'd also like to thank the dedicated bioinformatics community for their contributions to the extensive web of resources available.

# PREFACE

RNA sequencing data is very valuable in understanding the transcriptome. The goal of this book is to introduce readers to the fundamental concepts of RNA-seq experiments and guide them through the process of data analysis. This requires a thorough understanding of programming and genetics, so both of these areas will be explored in their individual chapters and combined when discussing RNA-seq.

This book covers the RNA-seq experimental design, sequencing library, quality control on raw data, alignment of reads to reference genome, read quantification, read count normalization, exploratory data analysis, differential gene expression analysis, and gene set enrichment analysis/pathway analysis.

Much more could be said about bioinformatics and RNA-seq, but this book serves as a brief introduction to the entire process, covering the steps of library preparation and sequencing all the way to pathway analysis.

## Who is this book for?

This book is intended for students and beginner researchers in computational biology. Since it covers both programming and basic biology, extensive background knowledge is not required. It is a great tool to help introduce wet lab scientists to computational methods.

RNA Sequencing: Principles and Data Analysis

# Table of Contents

*Bioinformatics* .................................................................. **1**

*R Programming* ............................................................. **4**

   **2.1 Introduction** .................................................... 4

   **2.2 Design** ............................................................ 5

   **2.3 Installation** ..................................................... 6
      2.3.1 Mac ................................................................... 6
      2.3.2 Windows ........................................................... 7

   **2.4 RStudio** .......................................................... 8

   **2.5 Objects** ........................................................... 9

   **2.6 Data Structures** ............................................ 10
      2.6.1 Vectors ............................................................. 10
      2.6.2 Matrices ........................................................... 12
      2.6.3 Data Frames ..................................................... 14
      2.6.4 Lists .................................................................. 15
      2.6.5 Factors ............................................................. 16

   **2.7 Packages** ...................................................... 17

   **2.8 Basic computations** ..................................... 18

   **2.9 Reading and Writing Data Files** ................... 19

*Genetics and Genomics* .............................................. **20**

   **3.1 DNA** ............................................................. 20
      3.1.1 Structure of DNA ............................................. 20
      3.1.2 Different Types of DNA .................................... 23

   **3.2 Replication** ................................................... 25

   **3.3 RNA** .............................................................. 28

3.3.1 Structure of RNA ................................................. 28
3.3.2 Different Types of RNA........................................ 30

## 3.4 Transcription and Translation ......................... 31
3.4.1 Transcription ........................................................ 32
3.4.2 Translation .......................................................... 35
3.4.3 Proteins ............................................................... 42

## 3.5 Gene Expression ............................................... 43
3.5.1 Transcriptional Regulation ................................... 43
3.5.2 Post-Transcriptional Regulation .......................... 46
3.5.3 Translational Regulation ..................................... 46

## 3.6 Mutations........................................................... 47

## 3.7 Genomics............................................................ 50
3.7.1 The Genome......................................................... 50
3.7.2 Human Genome Project....................................... 50

# *RNA Sequencing and Differential Gene Expression*.................................................. 52

## 4.1 Introduction ..................................................... 52

## 4.2 Experimental Design ........................................ 53

## 4.3 Sequencing Library .......................................... 54
4.3.1 Library Preparation – RNA Extraction ................. 54
4.3.2 Cluster Generation .............................................. 55
4.3.3 Sequencing.......................................................... 56

## 4.4 Raw Data ........................................................... 58
4.4.1 Output.................................................................. 58
4.4.2 Quality Control.................................................... 58

## 4.5 Aligning Reads to the Reference Genome ....... 59
4.5.1 Aligning with STAR .............................................. 60
4.5.2 SAM/BAM Files .................................................... 63
4.5.3 Integrative Genomics Viewer (IGV)...................... 64

## 4.6 Read Quantification.......................................... 66

## 4.7 Normalizing Read Counts ................... 67
### 4.7.1 DESeq2 ................................................. 68
### 4.7.2 edgeR ................................................... 72
## 4.8 Exploratory Data Analysis .................. 76
### 4.8.1 Principal Component Analysis (PCA) .............. 76
## 4.9 Differential Gene Expression Analysis .......... 84
### 4.9.1 Heatmaps ............................................. 87
### 4.9.2 MA Plot ................................................ 88
### 4.9.3 Volcano Plot ........................................ 89
### 4.9.4 Read Counts for Single Genes ............... 90
## 4.10 Gene Set Enrichment and Pathway Analysis .. 90
### 4.10.1 GSEA .................................................. 90
### 4.10.2 PANTHER ........................................... 92
### 4.10.3 Enrichr .............................................. 94
### 4.10.4 Fisher's Exact Test and Jaccard Index ........... 97

# | Chapter 1 |
# Bioinformatics

Bioinformatics is essentially the union of biology, computer science, and statistics. The advancement of technology has allowed scientists to generate massive datasets, especially in the field of genomics; The Human Genome Project is just one example. This interdisciplinary field was created from strong demand for data analysis and management techniques in biology and medicine. By definition, bioinformatics is the application of computational tools to understand and analyze biological data [1].

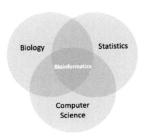

There are many applications of bioinformatics,

including the analysis of protein sequences, DNA sequences, RNA sequences, and entire genomes. This book will focus on RNA sequencing, shedding light on both the biology and the computer science principles involved.

Bioinformaticians are becoming increasingly valuable in research labs and companies all over the world. They can solve problems and create models on computers in a fraction of the time it would take using traditional methods. These individuals need to have a thorough understanding of biology *and* computer science, to the same extent that a general biologist or a general computer scientist would have in their respective fields. They need to specialize in two very complex fields and go beyond the standard curriculum to understand how to connect the two, which can be a deterrent for those considering the field. In a lab, bioinformaticians are arguably the translators between the wet lab researchers and their results. Therefore, bioinformaticians must be effective communicators; they must understand the biological experiments, conduct data analysis, and interpret the results (both biologically and statistically) to communicate back to the scientists.

The biology knowledge required for bioinformaticians can either be general (basics of molecular biology) or very specific to the experiments. The computer science background required includes a thorough understanding of machine learning, data mining, programming, data structures, and algorithms. Many bioinformaticians do not start in bioinformatics and instead come into the field with strong computer science backgrounds. This may seem daunting, but do not let it discourage you. There are many resources available for beginners and public datasets available to download and begin experimenting with. Since bioinformatics is a field that is constantly changing and has various applications, it is best to specify what you want to learn first and dive deep into that particular subject. In our case, that is RNA sequencing!

# | Chapter 2 |
# R Programming

## 2.1 Introduction

R is a dialect of the programming language S. S was developed by John Chambers at the Bell Laboratories in 1976 [1]. The current capabilities of R are continuously evolving with new versions being released annually. One important advantage of using R is its ability to create "publication quality" images and graphics. Visualization is key to genomic analysis and scientific literature because it is the only way to express complex data for the entire scientific community to understand. With R, the user can use a wide variety of existing tools/packages and even develop their own. The packages and functions available on R are based almost entirely on consumer demand and contributions from the community. If a package that you need does not exist, you need to create it yourself or employ someone to do it for you. However, the community is so extensive now that this problem is not regularly encountered. Arguably one of the most important parts of R is its large and active community in which programmers assist one another, create new packages, and share ideas. This is beneficial to both new and seasoned programmers, as there is information to assist anyone at any level of experience.

R is also free software, meaning that four freedoms are granted to the users, as stated by GNU (GNU publishes the primary source code for R).

1. The freedom to run the program as you wish, for any purpose (freedom 0).

2. The freedom to study how the program works, and change it so it does your computing as you wish (freedom 1). Access to the source code is a precondition for this.

3. The freedom to redistribute copies so you can help others (freedom 2).

4. The freedom to distribute copies of your modified versions to others (freedom 3). By doing this you can give the whole community a chance to benefit from your changes. Access to the source code is a precondition for this [2].

These freedoms make up the philosophy of using R. While they may seem trivial, it is important to understand when comparing it to other statistical packages.

## 2.2 Design

The R system has two basic "parts:" the first being the base R system and the second being "everything else," including over 4000 external packages. The base R system can be downloaded from CRAN with the following packages: base, utils, datasets, graphics, stats, grDevices, methods, grid, tools, parallel, tcltk, stats4, and compiler. For genomic data, Bioconductor offers a majority of the packages used in the analysis. Bioconductor is a free and open-source software project specifically created for genomic data.

## 2.3 Installation

You can install R on Linux, Mac, and Windows. The process is very simple.

### 2.3.1 Mac

1. Go to the following link: www.r-project.org
2. On the sidebar, there is a section called "Download." Click on the "CRAN" link here.
3. Select one of the CRAN locations listed under the country.

USA

| Link | Location |
|---|---|
| https://mirror.las.iastate.edu/CRAN/ | Iowa State University, Ames, IA |
| http://ftp.ussg.iu.edu/CRAN/ | Indiana University |
| https://rweb.crmda.ku.edu/cran/ | University of Kansas, Lawrence, KS |
| https://repo.miserver.it.umich.edu/cran/ | MBNI, University of Michigan, Ann Arbor, MI |
| http://cran.wustl.edu/ | Washington University, St. Louis, MO |
| http://archive.linux.duke.edu/cran/ | Duke University, Durham, NC |
| https://cran.case.edu/ | Case Western Reserve University, Cleveland, OH |
| https://ftp.osuosl.org/pub/cran/ | Oregon State University |
| http://lib.stat.cmu.edu/R/CRAN/ | Statlib, Carnegie Mellon University, Pittsburgh, PA |
| http://cran.mirrors.hoobly.com/ | Hoobly Classifieds, Pittsburgh, PA |
| https://mirrors.nics.utk.edu/cran/ | National Institute for Computational Sciences, Oak Ridge, TN |
| https://cran.revolutionanalytics.com/ | Revolution Analytics, Dallas, TX |

4. On the top of the page, click on "Download R for (Mac) OS X."

> **Download and Install R**
>
> Precompiled binary distributions of the base system and contributed packages, **Windows and Mac** users most likely want one of these versions of R:
>
> - Download R for Linux
> - Download R for (Mac) OS X
> - Download R for Windows
>
> R is part of many Linux distributions, you should check with your Linux package management system in addition to the link above.

5. Click on the file for the latest release of R.
6. Save and open the .pkg file. Follow the installation direction on-screen.

## 2.3.2 Windows

1. Go to the following link: www.r-project.org
2. On the sidebar, there is a section called "Download." Click on the "CRAN" link here.
3. Select one of the CRAN locations listed under the country.
4. On the top of the page, click on "Download R for Windows."
5. Click on "install R for the first time."

> **R for Windows**
>
> Subdirectories:
>
> base — Binaries for base distribution. This is what you want to **install R for the first time.**
> contrib — Binaries of contributed CRAN packages (for R >= 2.13.x; managed by Uwe Ligges). There is also information on third party software available for CRAN Windows services and corresponding environment and make variables.
> old contrib — Binaries of contributed CRAN packages for outdated versions of R (for R < 2.13.x; managed by Uwe Ligges).
> Rtools — Tools to build R and R packages. This is what you want to build your own packages on Windows, or to build R itself.
>
> Please do not submit binaries to CRAN. Package developers might want to contact Uwe Ligges directly in case of questions / suggestions related to Windows binaries.
>
> You may also want to read the R FAQ and R for Windows FAQ.
>
> Note: CRAN does some checks on these binaries for viruses, but cannot give guarantees. Use the normal precautions with downloaded executables.

6. Click on the "Download R 4.0.2 for Windows" at the top of the page (4.0.2 is the current version as of 6/22/2020). Save this file and run the .exe file.

7. Follow the on-screen installation instruction.

## 2.4 RStudio

RStudio is an Integrated Development Environment (IDE) for R. It is basically a more user-friendly platform for R, and it is open-source and free. You can download it at www.rstudio.com. There is both a desktop application (RStudio Desktop) and a server version that you can access through a web browser. The default pane layout consists of four main panels: the editor/source, the console, history/environment, and misc.

The editor is in the top left, and this is where code is saved and edited. The console is in the bottom left and it is where the commands are entered, and the output is printed for the user. The environment tab in the top right lists the loaded R objects, and the history tab in the same panel lists every console command in that project. Lastly, the misc. panel is in the bottom right and it has five tabs: files, plots, packages, help, viewer. Most of these are self-explanatory. The files tab is used to navigate the files in the specified folders. The plots tab contains the graphs that were generated. The packages tab contains the list of installed packages. The help tab acts as a search bar for the help directory, in addition to the output window for help commands in the console. The viewer tab is essentially a browser for local web content.

Figure 2.1 RStudio Desktop pane layout with the editor/source, console, history/environment, and misc.

## 2.5 Objects

R has six atomic classes of objects: character, numeric, complex, integer, logical (true/false), and raw (not generally used in data analysis). A vector is the most basic R object, as it can only contain objects that are part of the same class with the exception of a list. A list is a vector with objects of different classes [3].

R objects have attributes to describe the object. These include names, dimensions, class, and dimnames. They can be found with the attributes() function. The names attribute will label or index the elements of a vector. The dimensions attribute (dim) is used in arrays to return the size/length of the dimensions. The dimnames attribute labels each dimension independently to be used as headers. The class attribute controls and lists the classes that an object is derived from.

## 2.6 Data Structures

### 2.6.1 Vectors

Vectors are one of the fundamental data structures of R. They are lists of elements that all belong to the same class. Vectors can be created with the c() function.

```
> x <- c(1, 2, 3, 4, 5)
> x
[1] 1 2 3 4 5
```

Vectors can be used as an index. The vector index in R starts from 1, not 0. This means that the first element of a vector is designated with a 1, not a 0. Using the vector above as an example, the second element would be "2." The command x[2] would return the second element of that vector, which is "2." The command x[4] would return the fourth element of that vector, which is "4." The command x[c(1, 3)] would return the first and third elements of the vector, which are "1" and "3," respectively.

```
> x
[1] 1 2 3 4 5
> x[2]
[1] 2
> x[4]
[1] 4
> x[c(1, 3)]
[1] 1 3
```

Operations done on vectors will affect all elements of the vector. For example, if you add three to the vector x, each element in x will increase by 3. If you multiply the vector by 3, each element will be multiplied by 3. However, the original vector itself will not change with these operations.

```
> x
[1] 1 2 3 4 5
> x+3
[1] 4 5 6 7 8
> x*3
[1]  3  6  9 12 15
> x^3
[1]   1   8  27  64 125
```

To understand which class the elements of the vector belong to, the function class(x) can be used. The elements of x are numeric, as shown below.

```
> class(x)
[1] "numeric"
```

## 2.6.2 Matrices

Matrices are a group of data elements that are arranged in a two-dimensional rectangular shape. They have both rows and columns to represent the data. The data elements in the matrix must all be part of the same class. In the example below, m is being assigned the matrix of 4 rows and 2 columns. The output of m is an empty matrix with the assigned number of columns and rows. The dim(m) command returns the dimensions of the matrix. The contents of m get saved in the environment tab in the top right corner of the RStudio panel layout.

```
> m <- matrix(nrow = 4, ncol=2)
> m
     [,1] [,2]
[1,]  NA   NA
[2,]  NA   NA
[3,]  NA   NA
[4,]  NA   NA
> dim(m)
[1] 4 2
```

The image below shows the environment tab in the top right panel of RStudio. As you can see, the contents of m are defined here.

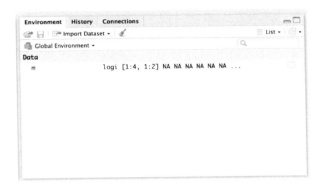

Matrices are made column-wise. This means that the entries begin in the upper-left corner of the matrix and run down the columns. The command below is organizing the numbers from one to ten in a matrix with 5 rows and 2 columns, as specified. The "1" is placed in the upper-left corner, the "2" is right below it (same column, different row), and so on.

```
> m <- matrix(1:10, nrow = 5, ncol = 2)
> m
     [,1] [,2]
[1,]    1    6
[2,]    2    7
[3,]    3    8
[4,]    4    9
[5,]    5   10
```

The functions rbind() and cbind() represent row-binding and column-binding, respectively.

```
> a <- 2:5
> b <- 21:28
> rbind(a, b)
  [,1] [,2] [,3] [,4] [,5] [,6] [,7] [,8]
a    2    3    4    5    2    3    4    5
b   21   22   23   24   25   26   27   28
> cbind(a, b)
     a  b
[1,] 2 21
[2,] 3 22
[3,] 4 23
[4,] 5 24
[5,] 2 25
[6,] 3 26
[7,] 4 27
[8,] 5 28
```

## 2.6.3 Data Frames

Data frames are similar to matrices in that they are used to store tabular data. Data frames are lists in which each element is the same length. An element of a list is comparable to a column and the length of the element is comparable to the number of rows. While matrices can only store elements of the same class, data frames can store elements of different classes in each column. Data frames also have row names and column names to provide more information about the columns and rows. They are created by reading in a dataset with the read.table() or read.csv() functions. Data frames can also be created with the data.frame() function. To convert a data frame to a matrix, the function data.matrix() can be used.

## 2.6.4 Lists

Lists are vectors that can have elements belonging to different classes. They can be created with the list() function.

```
> a <- list(number = c(2, 4, 6),
+           matrix = matrix(1:6, ncol = 3, nrow = 2),
+           name = "Tom")
> a
$number
[1] 2 4 6

$matrix
     [,1] [,2] [,3]
[1,]    1    3    5
[2,]    2    4    6

$name
[1] "Tom"
```

To find an element of the list (in this example: a), the a[] function can be used. The number between the brackets represents the position or name of the element in the list, and "a" is the name of the list. For example, a[3] would output the element of the list that is in the third position. The function a["matrix"] would output the element of the list that is named "matrix," which is the 2x3 matrix defined in the earlier step.

```
> a
$number
[1] 2 4 6

$matrix
     [,1] [,2] [,3]
[1,]   1    3    5
[2,]   2    4    6

$name
[1] "Tom"

> a[3]
$name
[1] "Tom"

> a[1]
$number
[1] 2 4 6

> a["matrix"]
$matrix
     [,1] [,2] [,3]
[1,]   1    3    5
[2,]   2    4    6
```

## 2.6.5 Factors

Factors represent categorical data. They are very important in statistical modeling, as each element has a label. Factors can either be ordered or unordered depending on the situation. They can be created with the factor() function and work with modeling functions like lm() and glm(). The unclass() function shows the representation and levels of the factor, being "female" and "male" in this example.

```
> w <- factor(c("male", "female", "male", "female", "female"))
> w
[1] male   female male   female female
Levels: female male
> table(w)
w
female   male
     3      2
> unclass(w)
[1] 2 1 2 1 1
attr(,"levels")
[1] "female" "male"
```

## 2.7 Packages

R packages are one of the fundamental units of R. They are collections of functions and data that carry out operations and tasks vital to certain programs. The majority of packages available for R come from the Comprehensive R Archive Network, or CRAN. As of 2020, the CRAN package repository has just over 16,000 packages available [4]. They are developed by the community, which is important in the success of R and its packages; if you have a problem, there is a great chance that someone else has already encountered it and has a package available to solve the issue that you can download. This ensures that individuals do not spend time repeating solutions to the same problems.

Packages from CRAN can be installed with install.packages("x"). You can use these packages in R with library("x"). CRAN packages can be updated with update.packages().

CRAN is a repository for R packages, meaning that it is a central location where developed packages are available for download. Bioconductor is a repository that is specific to bioinformatics (analysis of genomic data and statistics). These packages must be installed through Bioconductor. To do this, the BiocManager package must be installed.

>     install.packages("BiocManager")
>
>     BiocManager::install("DESeq2")

GitHub is another repository that is not R-specific but includes many R packages. These packages must be installed through GitHub.

>     library(devtools)
>
>     install_github("x")

## 2.8 Basic computations

R can be utilized as a calculator, as many complex functions are built off of these fundamental mathematical concepts.

```
1 + 3            # Simple addition
4 * 2            # Simple multiplication
10 / 5           # Simple division
16 + 4 * 1       # Order of operations taken into account
2^3              # 2 raised to the third power
log(10)          # Natural logarithm with base e
sqrt(25)         # Square root
abs(1-8)         # Absolute value
pi               # The number pi (3.14159...)
```

## 2.9 Reading and Writing Data Files

Data files can be read into R with a variety of functions, many specific to the types of data files and their origins.

| Reading Data Files | |
|---|---|
| read.table() | Reading tab-/space-delimited tabular data |
| read.csv() | Reading comma-delimited tabular data |
| read.fwf() | Reading fixed-width data |
| load() | Reading data in saved workspaces/format |
| source | Reading in R code files from file/connection/URL |
| dget | Reading in R code files with ASCII text representation |
| unserialize | Reading R objects in binary format |
| readLines() | Reading lines of text file |

You can also write data files in R, using the function specific to the type of data.

| Writing Data Files | |
|---|---|
| write.table() | Writes tabular data to connection/text file |
| writeLines() | Writes data line-by-line to connection/file |
| dput | Outputs a textual representation of the object into the connection/file (object name not written) |
| dump | Outputs a textual representation of the objects on a connection/file |
| save | Writes the representation of objects in a binary format on the connection/file. |
| serialize | Converts objects to their binary format to output to a connection/file. |
| write.table() | Writes tabular data to connection/text file |

# Chapter 3
# Genetics and Genomics

## 3.1 DNA

DNA, or deoxyribonucleic acid, is the basis of every living thing (and some viruses) on Earth. DNA stores genetic information and passes it on through generations. Its shape allows for self-replication, which is very important in cell division.

### 3.1.1 Structure of DNA

You've probably seen DNA represented as a twisted ladder. This double helix structure of DNA is what makes many of its processes possible and simplifies the mechanisms involved. The DNA in just one cell would be over six feet long if stretched out. Now, multiply this by 100 trillion, which is the number of cells in the human body. The coiled structure of DNA is the reason it is able to fit inside cells.

Nucleosomes are formed when the DNA is coiled around proteins. To hold these coils together and keep their shape, proteins called histones are employed [1].

Chromosomes are groups of packed DNA in the nucleus. A full set of chromosomes, 23 pairs for a human, are stored in the nucleus of all eukaryotes. This information is the blueprint for building the organism, as the instructions are given in the form of genes. The genes are sections of the DNA and make up the chromosomes, which code for physical traits, functions, etc.

DNA's structure is derived from three basic components: phosphate groups, deoxyribose sugars, and nitrogen bases. A nucleotide is the combination of a nitrogen base, phosphate, and sugar. There are four nitrogenous bases in DNA: Adenine (A), Thymine (T), Cytosine (C), and Guanine (G). These four bases belong to two main categories: purines and pyrimidines. Both adenine and guanine are purines since their structures have two rings, while both thymine and cytosine are pyrimidines because their structures have a single, six-sided ring. These bases are "flat" molecules (due to ring structure), giving them the ability to stack up, which increases the strength of DNA and decreases the size [1].

| Pyrimidines | | Purines | |
|---|---|---|---|
| Cytosine | Thymine | Adenine | Guanine |

These nucleotides are bonded together with deoxyribose sugar and a phosphate molecule. The deoxyribose sugar is derived from the ribose produced when the body breaks down ATP (the energy unit of the body). Deoxyribose lacks an oxygen atom compared to the ribose, hence the name DEOXYribose.

We can't discuss the structure of DNA without mentioning Rosalind Franklin, James D. Watson, Francis Crick, and Erwin Chargaff. An experiment in 1952, called the Hershey-Chase experiment, showed that DNA was the material that stored genetic information. Scientists around the world began racing to uncover the three-dimensional structure of DNA. James D. Watson, 23 years old at the time, and had traveled to Cambridge University to work alongside Francis Crick, who was using X-ray crystallography to study protein structure. They went to King's College to visit the laboratory of Maurice Wilkins, where Rosalind Franklin was working. Watson saw her X-ray image of DNA and used it to make his big discovery. He understood that the basic shape of DNA was a spiral with a uniform diameter and that the nitrogenous bases were stacked on top of one another. He also found that the thickness of this spiral meant that it was made up of two polynucleotide strands, making it a double helix. Franklin had concluded that the nitrogenous bases were on the inside of the molecule, while the sugar-phosphate backbone was on the outside of the molecule, so Watson and Crick used this and the X-ray image as a foundation to evaluate the base-pairing rules. They first assumed that bases pair with the same bases, meaning A paired with A and G paired with G, but this did not support the uniform diameter found in the X-ray image, since the sizes of these same-base pairs would vary greatly. This led them to the conclusion that a purine (double-ring base) must pair with the pyrimidine (single-ring base). This was the only way for the sizes of the pairs to abide by the uniform diameter principle. After further study, they found that A pairs with T and C pairs with G based on optimal hydrogen bonding. Erwin Chargaff, years earlier, had deduced that the amount of adenine in DNA was equal to the amount of thymine, and the amount of cytosine was equal to the amount of guanine. Watson and Crick's conclusions support Chargaff's discovery. Watson and Crick published their paper in the journal *Nature* for the molecular model of DNA in 1953 and Watson, Crick, and Wilkins received the Nobel Prize in 1962. Unfortunately, Rosalind Franklin passed away from ovarian cancer in 1958, causing her to be disqualified from receiving the honor [3].

While we often credit Watson and Crick for the discovery of the double helix shape of DNA, it is important to recognize Rosalind Franklin for her immense contribution. Without her X-ray image, the discovery of DNA's structure would have been delayed for many years.

### 3.1.2 Different Types of DNA

**I. Nuclear DNA**

Nuclear DNA is DNA found in the nucleus of the cell. It codes for the physical traits of an organism, which is called its phenotype. It is packed up in chromosomes and passed from the parents to the offspring through sexual reproduction. Human nuclear DNA is involved in the sequencing of the human genome.

## II. Mitochondrial DNA

The mitochondria, commonly defined as the "powerhouse of the cell", is present in all animals, fungi, and plants. In the mitochondria, you can find mitochondrial DNA (mtDNA). The shape and size of mtDNA differ significantly from that of nuclear DNA. While nuclear DNA is linear and longer, mtDNA is circular and much shorter, with exactly 37 genes in the entire molecule and fewer than 17,000 base pairs. While DNA is inherited from both parents, mtDNA only comes from the mother since it is passed through the egg cell's cytoplasm. Cellular metabolism is controlled by mtDNA, which is the process in which chemical energy from food is converted into adenosine triphosphate (ATP) [4].

## III. Chloroplast DNA

Chloroplasts are organelles in plants used in photosynthesis to convert light to chemical energy. Chloroplast DNA molecules are circular, similar to the shape of mtDNA, but are much larger with about 120,000-160,000 base pairs and 120 genes. Chloroplast DNA can come from either the mother or the father and the genes are generally used to guide the processes of photosynthesis [5].

## 3.2 Replication

Replication is the process of copying DNA during cell division. The double helix structure of DNA is what allows for this process to occur. Watson and Crick inferred that a cell uses the complementary bases when copying its DNA, and they were correct. The two strands of DNA in the double helix act as a template in forming a new strand. The free nucleotides in the nucleus link to the template strand according to classic base-pairing rules. This creates a new DNA strand complementary to the "parent" strand. Watson and Crick's inference stated that when the DNA undergoes replication, each new DNA molecule will have one strand from the parent DNA and one new complementary strand. This model is called the semiconservative model, and it was proven to be true in the 1950s. DNA replication is based on this fundamental concept.

The origin of replication is a segment of DNA that has a specific set of nucleotides. Proteins attach to the origin of replication on the DNA and separate the two strands. There are thousands of origins of replication on the DNA molecule, decreasing the time necessary for replication to occur. The process of replication proceeds to the left and the right of the origin of replication, which creates "bubbles," as shown in the figure. The daughter strands begin to form and are complete once the thousands of bubble regions merge. The final product is two daughter DNA molecules.

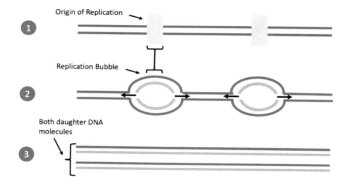

Figure 3.1 **DNA replication bubbles.** (1) The parent DNA double-stranded molecule has multiple origins of replication where replication takes place. (2) The parental DNA opens up so that the daughter strands can begin to form abiding by classic base-pairing rules. These different origins of replication eventually merge. (3) The product of replication is two daughter DNA molecules.

While this is seemingly simple, replication must follow a set of biological rules. The sugar-phosphate backbone of the parent DNA's strands run in opposite directions; the strand has a 3' and a 5' end (3' is stated "three-prime" and 5' is stated "five-prime"). The 3' end has a hydroxyl group and the 5' end has a phosphate group. DNA polymerase can only add nucleotides to the 3' end, meaning that the daughter DNA strand can only be synthesized in the 5' → 3' direction. While one daughter strand can be synthesized continuously, the other daughter strand must be synthesized in short pieces that still abide by the 5' → 3' direction requirement. As the fork in the replication "bubble" opens up, the new strand must be synthesized in small pieces, called Okazaki fragments, that are linked with

DNA ligase. Both DNA ligase and DNA polymerase are enzymes that are essential to this process.

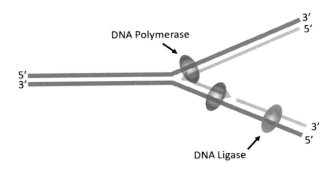

Figure 3.2 **Simplified depiction of one side of the replication bubble.** DNA polymerases (shown in black) can only add nucleotides to the 3' end, so DNA is synthesized in the 5' → 3' direction. One of the daughter strands can be synthesized continuously according to these rules (the top strand in this case), whereas the other strand must be synthesized in pieces (called Okazaki fragments) in the opposite direction. These fragments are linked together with DNA ligase (shown in green).

DNA polymerase has another role in addition to adding the nucleotides: proof-reading. DNA polymerase proofreads the new strands and removes nucleotides that have paired incorrectly. Proofreading is very important because the DNA replication speed in humans is about 2,000 bases per minute. If a mistake is recognized, the DNA polymerase cuts out the wrong nucleotide base in a process called exonuclease activity in which the DNA polymerase moves in 3' → 5' direction (the opposite of synthesis). The error rate after proofreading is about one in 10 million base pairs. Additionally, both DNA polymerase and DNA ligase repair damaged DNA that results from radiation or toxic chemicals. To finalize replication, the DNA is wrapped around nucleosomes, which are structures that allow DNA to fit into the nucleus. Supercoiling is this process of twisting and wrapping the DNA. About 150 base pairs of DNA are wrapped around each nucleosome and kept in place with histones. Before replication, the DNA must be removed from the nucleosomes for the enzymes to work properly.

Replication is a seemingly simple, yet very complex, process. It is crucial to an organism's development and sexual reproduction. The entire purpose of replication is to guarantee that each cell in an organism carries the same genetic information and can be passed onto offspring [6].

## 3.3 RNA

### 3.3.1 Structure of RNA

Ribonucleic acid (RNA) works with DNA to carry the messages from the nucleus of the cell to the cytoplasm and direct the process of translation. RNA is composed of a ribose sugar, four nucleotide bases, and a phosphate group. It is similar to the makeup of DNA, except RNA uses ribose sugar while DNA uses deoxyribose sugar, RNA is single-stranded, RNA uses uracil instead of thymine [7].

The two sugars differ in their chemical structure: ribose has an OH at both the 3' and 2' ends whereas deoxyribose has an OH on the 3' end and just an H on the 2' end. The reason for this is to protect the DNA from being unstable and decomposing. The OH group is called the reactive group because of the oxygen atom is chemically aggressive. The fact that DNA is missing an oxygen atom helps it avoid being involved in chemical reactions and ultimately getting broken down.

The four bases of RNA are adenine (A), guanine (G), cytosine (C), and uracil (U). As you may notice, this is different from DNA's bases regarding one nucleotide; RNA has uracil (U) instead of thymine (T). Uracil is actually the precursor to DNA because it works with folic acid to add a methyl group to uracil, which results in thymine. While these are two separate bases, they both pair with adenine (A). The reason for the difference is that thymine has a methyl group to ensure that nucleases can't attach to DNA and break the bonds. Uracil is less specific than thymine in regard to what it can form pairs with, which is beneficial to RNA due to its single-stranded nature and structure.

RNA is single-stranded, which is its primary structure, but it also has a secondary structure. This is a result of RNA bonding to itself and folding up.

| Pyrimidines | | Purines | |
|---|---|---|---|
| Cytosine | Uracil | Adenine | Guanine |

## 3.3.2 Different Types of RNA

### I. Messenger RNA (mRNA)

Messenger RNA (mRNA) is the single-stranded RNA molecule that comes from transcription (converting the genetic message of DNA to mRNA). It is complementary to the template strand. After transcription, mRNA moves out of the nucleus and into the cytoplasm for translation. The mRNA works with the ribosomes during translation, as the ribosome connects the mRNA to the tRNA to translate the genetic information to amino acids.

### II. Transfer RNA (tRNA)

Transfer RNA (tRNA) helps translate the message of the codons on the mRNA to the amino acids. It has the ability to recognize the correct codons in the mRNA that corresponds with the amino acids. The structure of tRNA is essentially a single RNA strand that folds into itself to form many double-stranded regions (connected via hydrogen bonds). The tRNA has a single-stranded region at one end called the anticodon. The anticodon has three bases that are complementary to a codon on the mRNA. The acceptor arm is the tail on the tRNA that the amino acid attaches to [8].

### III. Ribosomal RNA (rRNA)

Ribosomal RNA (rRNA) is a non-coding RNA that is found in ribosomes. It aids in protein by putting the amino acids together to form a polypeptide chain.

These rRNA molecules also bind the tRNAs and other molecules that are involved in translation [9].

## 3.4 Transcription and Translation

An organism's genotype is its genetic makeup and its phenotype is its physical traits. The organism's DNA makes up its genotype, and the proteins control its phenotype. The link between the genotype and phenotype is that the organism's DNA dictates protein synthesis. This process can be simplified: DNA → RNA → Protein. Transcription is DNA → RNA, in which RNA is synthesized as dictated by the DNA. Translation is RNA → Protein, in which proteins are synthesized as dictated by the RNA.

This link was first proposed by Archibald Garrod in 1909. Garrod was an English physician who believed that genes dictate phenotypes with enzymes. He used alkaptonuria as an example, which is a hereditary condition where the urine is a darker color because it contains alkapton. His theory was that most people had the enzyme to break down this chemical, but those with alkaptonuria did not inherit this enzyme, and therefore could not break down the alkapton. Later, in the 1940s, George Beadle and Edward Tatum used the bread mold *Neurospora crassa* to demonstrate this relationship. They studied the strains of the bread mold that could not grow in the normal growth medium and found that they did not have a certain enzyme in the pathway that synthesized the amino acid was involved in the mold's growth. They also found that each one of these strains was lacking or defective in a gene. They hypothesized that a gene dictates the synthesis of an enzyme. Since this discovery, this hypothesis has been adapted to include all polypeptides [10].

### 3.4.1 Transcription

Transcription (DNA → RNA) occurs in the nucleus. There are three main steps: initiation, elongation, and termination [11].

#### 3.4.1.1 Initiation

Human chromosomes have approximately three billion base pairs of DNA, but only about 1% of DNA actually gets transcribed into mRNA [12]. To begin transcription, the genes must first be located. On the DNA, there are regions called promoters and terminators. The promoter is the sequence on the DNA that tells the enzymes where to begin and is generally about 30 base pairs from the genes of interest. In eukaryotes, the beginning of the promoter sequence is always TATA. This is an example of a consensus sequence, meaning that the sequence has the same function wherever it is found.

Since DNA is double-stranded, one strand is the nontemplate strand and the other is the template strand. Even though they are complements of one another, they will code for different mRNA, and therefore different proteins. For example, if a region on one DNA strand is ACCTTG, the bases on the other strand would be TGGAAC. The first nucleotide sequence, when converted to mRNA, would be UGGAAC. The second nucleotide sequence would be

ACCUUG. The first mRNA sequence, when converted into amino acids, would be Trp-Asn whereas the second mRNA sequence would be Thr-Leu. These are two different amino acid sequences, which shows how important it is to understand which strand is the template strand and which is the nontemplate strand. The template strand is the DNA strand that transcription is being based on, and the nontemplate strand is the DNA strand that is the original message being transcribed. The promoter region is on the nontemplate strand, indicating that the other strand is the template strand for the RNA polymerase. RNA polymerase can recognize the bases on the DNA and add the complementary bases to the mRNA strand by working with holoenzymes.

To start the initiation stage, the RNA polymerase finds the promoter region of the DNA and separates the two strands of DNA. It does so by breaking the hydrogen bonds between the two strands, creating a "transcription bubble." The RNA polymerase reads the template strand and creates the mRNA strand with the ribonucleotides complementary to the template strand on the 3' end. The new RNA strand is separated from the DNA. After the RNA polymerase is done with a section of DNA, the two strands DNA bind together again.

3.4.1.2 Elongation

The RNA polymerase keeps working through the template strand, adding to the new RNA strand. This keeps going until the entire unit is transcribed. After the RNA polymerase is finished with a certain section (about 20 base pairs), binds together the two separated DNA strands and moves to the next section. Within the DNA of eukaryotes, there are introns and exons. Introns are sequences that do not code for any traits but play a role in gene expression. Exons are sequences that code for traits. In transcription, both the introns and the exons are transcribed, but the coded introns in the RNA get removed in later steps. Prokaryotes have no introns.

### 3.4.1.3 Termination

The elongation process continues until the RNA polymerase reaches the terminator. Like the promoter, the terminator is a region on the DNA that tells the RNA polymerase to end the transcription process. In prokaryotes, the terminator causes the new RNA strand to fold back on itself. In eukaryotes, a protein called the termination factor helps the RNA to know when to stop transcribing. The DNA molecule returns to its original double helix structure. The mRNA must go through a preparation process before translation, which includes adding units and editing. A 5' cap must be added onto the mRNA molecule so it can be recognized by the ribosome during translation. One of the three phosphates from the end of the mRNA strand is removed, allowing guanine to take its place. Multiple methyl groups attach to the guanine and the first two nucleotides on the mRNA. A string of adenine bases, called the poly-A tail, must be added to the 3' end of the mRNA strand in eukaryotes to protect it from decomposition. The length of this string (50-250 A) determines how long the mRNA lasts before decomposition, as the mRNA is very easily degraded. After this, the aforementioned introns must be removed from the mRNA. This process is called RNA splicing, and it is when the introns are cut out from the sequence and the exons are pulled together. The spliceosome, a complex of proteins that aids the process, pulls the two ends of the intron together to form a loop and then cuts out the intron so that the ends of the two exons are together. A phosphodiester bond is created between the two exons [12].

After the mRNA is prepped, it can move out of the nucleus and into the cytoplasm from translation.

### 3.4.2 Translation

Translation (RNA $\rightarrow$ proteins) occurs in the cytoplasm with three main steps: initiation, elongation, and termination [13].

### 3.4.2.1 Codons

After discovering that DNA was the genetic material, scientists debated how those four nucleotide bases of A, U, C, and G could code for 20 amino acids. If one base coded for one amino acid, there would only be four amino acids possible. If two bases coded for one amino acid, there would only be 16 amino acids possible (AA, AU, AC, AG, UA, UU, UC, UG, etc.). If three bases code for one amino acid, there would be 64 possible combinations for amino acids. This meant that three bases could only code for one amino acid, but one amino acid could be encoded by multiple combinations of three bases. The code is redundant but never ambiguous. This code is referred to as the triplet code. This redundancy can be beneficial because it makes the code more flexible and leaves some room for error. Out of the 20 amino acids, 18 amino acids have more than one codon coding for them. The genetic code is nearly universal, as almost every organism on the planet interprets the code in the same manner. This is important because it allows scientists to study and swap genes across species. This also means that the genetic code evolved very early in history, as it is shared amongst most modern organisms.

There are 64 total codons, representing every possible combination of the bases A, U, C, and G. Three of these codons do not code for an amino acid. Their primary role is to signal "stop" to the ribosome in translation. These codons are UAA, UAG, and UGA. One specific codon (AUG) codes for the amino acid methionine *or* signals "start." It is important to remember these start and stop codons, as they are vital in the initiation and termination stages of translation [14].

Reading the code is simple. Each sequence starts with the "start" codon, AUG. After the start codon, the sequence is read in groups of three until a stop codon (UAA, UAG, UGA) is detected. Each triplet corresponds to an amino acid on the codon chart and these amino acids are written with dashes in between each one.

Second Base of Codon

| | | U | C | A | G | |
|---|---|---|---|---|---|---|
| First Base of Codon | U | UUU, UUC Phe<br>UUA, UUG Leu | UCU, UCC, UCA, UCG Ser | UAU, UAC Tyr<br>UAA Stop<br>UAG Stop | UGU, UGC Cys<br>UGA Stop<br>UGG Trp | U<br>C<br>A<br>G |
| | C | CUU, CUC, CUA, CUG Leu | CCU, CCC, CCA, CCG Pro | CAU, CAC His<br>CAA, CAG Gln | CGU, CGC, CGA, CGG Arg | U<br>C<br>A<br>G |
| | A | AUU, AUC, AUA Ile<br>AUG Met or START | ACU, ACC, ACA, ACG Thr | AAU, AAC Asn<br>AAA, AAG Lys | AGU, AGC Ser<br>AGA, AGG Arg | U<br>C<br>A<br>G |
| | G | GUU, GUC, GUA, GUG Val | GCU, GCC, GCA, GCG Ala | GAU, GAC Asp<br>GAA, GAG Glu | GGU, GGC, GGA, GGG Gly | U<br>C<br>A<br>G |

Third Base of Codon

Figure 3.3 Codon chart for amino acids.

## 3.4.2.2 Transfer RNA (tRNA)

Transfer RNA (tRNA) is the translator between codons and amino acids. The function of tRNA is analogous to a translator (who speaks both languages) in a meeting where the two government officials do not speak the same language. The tRNA molecules must be able to pick up the correct amino acids *and* recognize the correct codons in the mRNA. Its structure allows for it to do both. The structure of tRNA can be depicted in two ways. It is made up of a single RNA strand that twists and folds onto itself to form many double-stranded regions that pair with hydrogen bonds. The single-stranded loop on one side of the folded tRNA is called the anticodon. The anticodon has three bases that are complementary to a specific codon on the mRNA. The acceptor arm is the single-stranded tail on the tRNA which is the attachment site for the amino acid. The tRNA molecules, while shown in their simplified structures, vary slightly based on the amino acid. The amino acid is connected to its specific tRNA with a unique enzyme. Since there are 20 amino acids, there are 20 enzymes. One molecule of ATP is utilized in this binding process [15].

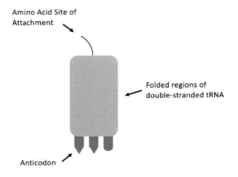

Figure 3.4 **Simplified depiction of tRNA.** The amino acid attachment site is the single-stranded region of the tRNA and the anticodon is the single-stranded loop at one end of the fold. Both of these components work together to ensure that tRNA can match the codon on the mRNA with the corresponding protein.

### 3.4.2.3 Ribosomes

Ribosomes are vital in translation since they are the structures that coordinate the functions of mRNA and tRNA to create polypeptide chains. A ribosome has two subunits: a large subunit and a small subunit. Both of these units are made up of ribosomal RNA (rRNA). The ribosome holds the mRNA and tRNA together and allows for complementary base pairing between the two. It connects the amino acids from the tRNA to the polypeptide chain. There are two sites on the ribosome: A-site and P-site. The A-site is the acceptor site where the tRNA molecules place their anticodon to pair with the codon from the mRNA. The P-site is the peptidyl site where the amino acids get tied to one another with peptide bonds [16].

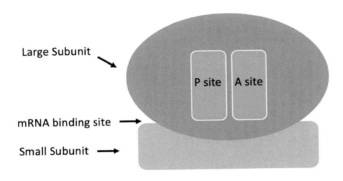

Figure 3.5 **Structure of ribosome with small ribosomal subunit and large ribosomal subunit.** The tRNA attaches to the P site and A site and the mRNA binds to the region between both subunits to interact with the tRNA.

### 3.4.2.4 Initiation

Like transcription, translation has three main steps: initiation, elongation, and termination. The mRNA molecule from transcription has a 5' cap with a region of nucleotides following it that is not part of the genetic message, and a 3' tail with a region of nucleotides before it that is also not part of the genetic message. This is important because it allows the mRNA to bind to the ribosome and ensure that translation begins at the desired location. The mRNA binds to the small subunit on the ribosome and the initiator tRNA pairs with the start codon on the mRNA molecule. The tRNA has the amino acid methionine. If you recall, methionine is the amino acid that corresponds to the start codon, AUG, on the codon chart. The anticodon on the tRNA is UAC, which is the complementary triplet to the start codon. The start codon and the anticodon base-pairs. After this, the large subunit binds to the small unit, resulting in a "functional ribosome." The initiator tRNA gets placed into the P-site. At this point, the A-site is empty and waiting for the next tRNA with its amino acid [17].


### 3.4.2.5 Elongation

The general goal of elongation is to add amino acids to the polypeptide chain. The process has three key components: codon recognition, peptide bonding, and translocation. Codon recognition is when the anticodon of the tRNA pairs with the mRNA codon in the ribosome's A site. During peptide bonding, the existing polypeptide chain separates from the new tRNA (in the P site) and bonds to the amino acid on the new tRNA in the A site. The A peptide bond forms, which results in the addition of a new amino acid to the existing polypeptide chain. The last step is translocation. Once the amino acids are linked to one another, the tRNA in the P site leaves the ribosome.

Remember, the tRNA in the P site no longer has any amino acids attached to it. The tRNA in the A site (which has the polypeptide chain) moves from the A site to the P site. Since the tRNA is hydrogen-bonded to the mRNA through their anticodon/codon, the tRNA and mRNA move together. These three steps keep occurring from the 5' to 3' direction of the mRNA until the stop codon is reached in the A site of the ribosome. The stop codons are UAA, UAG, and UGA [18].

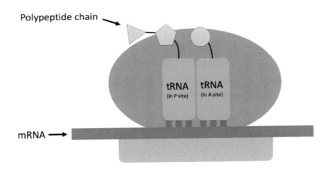

Figure 3.6 Activated ribosome with tRNA in both the P site and A site; growing polypeptide chain during elongation.

### 3.4.2.6 Termination

When the A site of the ribosome detects the stop codon (UAA, UAG, or UGA), elongation ends. At this point, no more tRNAs enter the A site of the ribosome, and the tRNA (with its completed polypeptide chain) is in the P site. Proteins called release factors bind to the ribosome and recognize the stop codon to cleave the polypeptide chain from the tRNA. The ribosome then detaches from the last rRNA and the subunits separate to find another mRNA. The entire process is very rapid, as mRNA can be translated multiple times and by more than one ribosome at a time.

### 3.4.3 Proteins

Proteins are the second most common substance in cells. Structure determines function for proteins. Each one of the 20 amino acids has a positively charged amino group ($NH_2$) and a negatively charged carboxyl group (COOH) attached to the central carbon atom. The amino group and the carboxyl group are always opposite to each other in regard to the central carbon atom. The atoms that form the branches/rings of these structures, called radical groups, are unique to the amino acid. There are five main radical groups: hydrophilic, hydrophobic, positively charged (acids), negatively charged (bases), and aromatic R groups. These groups, and their compatibility with one another, are what determines their shape (they either attract or repel other groups).

There are four levels of protein structure: primary, secondary, tertiary, and quaternary. The primary structure is the simple, unfolded structure of the amino acids in the polypeptide chain. The secondary structure is the folded structure based on the interactions between atoms of the polypeptide chain. The alpha-helix and beta- pleated sheet are the two most common secondary structures. The tertiary structure is the overall three-dimensional structure of the polypeptide chain, based on the interactions between the R groups. The quaternary structure applies to proteins that are made up of multiple polypeptide chains (subunits), as this level represents the combination of these subunits.

Proteins may lose their structure if exposed to certain environmental conditions, such as extreme heat, chemicals, temperature change, or pH change. In this case, the protein would lose its three-dimensional structure and revert to the string of amino acids. This is called a denatured protein. Since a protein's structure defines its function, denatured proteins generally do not function properly or at all. Some proteins can reverse the process of denaturation by folding back into its functional form. While this is possible for some proteins, it is not true for all. For example, denaturation can't be reversed in the context of frying an egg. The egg white has a protein called albumin that turns solid and opaque when denatured by the stove's heat. At least from my own experience, I know that the egg white does not turn back to its liquid form when it cools down. This means that the process of denaturation is permanent, as the proteins cannot fold back into their pre-denaturation forms [19].

## 3.5 Gene Expression

### 3.5.1 Transcriptional Regulation

The gene regulation in eukaryotes occurs during transcription. If a gene is considered to be "on," it is being transcribed. Likewise, if a gene is considered to be "off," it is not being transcribed. Gene expression is entirely dependent on if and how transcription occurs since transcription results in an mRNA strand that can be translated into a protein and therefore become expressed. Gene expression is controlled by the timing of transcription, which is dependent on regulation factors (enhancers), cell signals with hormones, and DNA accessibility. This can also include histone modifications and DNA methylation.

DNA accessibility is important in transcription because the DNA is very tightly coiled to fit into the nucleus and cannot be transcribed without slightly unwinding. This tight packaging keeps genes in their default "off" position; genes are off by default to keep the body in order and ensure that genes are only turned on when necessary in specific tissues. The tight coils prevent transcription from occurring because the transcription factors cannot access the genes. Another way that genes remain off is through repressors, which are proteins that bind to the same DNA sites that the transcription factors would have normally tried to bind to. This prevents the unwinding of DNA, which prevents transcription. Some regions of DNA are prepackaged to readily unwind and ease the beginning of transcription. Genes can be turned on by removing the DNA from their tight coils with the use of specific proteins called chromatin-remodeling complexes. They attach to the region of DNA and push the histones away, which invites transcription facts to begin the process of transcription [20].

There also exists regulation from other genes, including enhancers, silencers, insulators, and transposable elements [21]. Enhancers start transcription and speed up the entire process. They are tissue-specific, meaning that they can only direct genes that are usually activated in the cell type. Enhancers work alongside transcription factors and can influence genes that are over several kilobases away. This is possible because the enhancers cause the DNA to loop around. Silencers are sequences that work with repressor proteins to slow down or completely stop transcription by keeping the DNA tightly coiled. Silencers can also be several kilobases away from the target genes and cause the DNA to loop around. Insulators, also called boundary elements, can create a barrier between non-target DNA regions and an enhancer/silencer. This ensures that the effect of the enhancer/silencer is confined to the intended region. Lastly, transposable elements are genes that move around and can cause mutations and regulate other genes. About 50% of human DNA is made up of transposable elements. There are very specific parameters for where the transposable elements can insert themselves in the DNA [20, 21].

Hormones can also control gene expression. They are chemicals that circulate in the bloodstream and can affect tissues throughout the body. Small and fat-soluble hormones can pass into the cell directly to find receptor proteins and form a complex that acts as a transcription factor for certain genes. However, larger hormones cannot pass into the cell directly, so they use receptor proteins in the cell to convey their message. This process is called signal transduction. Hormone response elements (HREs) are the genes that react to these sequences in order to influence the genes and the rate of transcription for that gene [21].

## 3.5.2 Post-Transcriptional Regulation

Splicing is the process of removing non-coding introns from the mRNA after transcription. Multiple combinations of exons can result from splicing, and exons can be edited out as well. This allows for one gene to code for more than one protein. This genetic flexibility is dictated by splicing regulators on both the exons and the introns. These regulators include exonic splicing enhancers, exonic splicing silencers, intronic splicing enhancers, and intronic splicing silencers. When a repressor protein binds to the strand, it reduces the chance of that site being used as a splice junction. On the other hand, when an activator protein binds to the strand, it increases the chance of that region being used as a splice junction.

Currently, researchers are studying mRNA silencing as a method of regulation. The mechanisms for this RNA interference are largely unknown, but it is important in regulating gene expression and protecting organisms from genes of viruses.

Noncoding RNA (ncRNA) is also being researched regarding its regulatory roles. Examples of ncRNAs produced are long non-coding RNAs (lncRNAs), small-interfering RNAs (siRNAs), microRNAs (miRNAs), promoter-associated RNAs, and small nucleolar RNAs (snoRNAs) [21, 22].

## 3.5.3 Translational Regulation

There are three ways to regulate gene expression during translation: changing *where* translation happens, changing *when* translation happens, and altering the protein shape.

Modifying where the mRNA is found in the cytoplasm can help with gene control. The mRNA can be limited to a specific section of the cytoplasm so that the proteins from translation would only be present in that specific section.

Controlling the timing of translation also aids in gene control. All mRNAs have a region of nucleotides on their 5' end that is not translated. The length and contents of this sequence control the timing of translation, with some cells depending on the conditions of their environment (ex. chemicals such as hormones).

Translation produces proteins, which directly dictate gene expression. The protein's shape and components are very important, and gene expression is dependent on both factors. Components, such as phosphates or carbohydrate chains, can be added to the protein, which would change the original function of the protein and result in altered gene expression [21, 23].

## 3.6 Mutations

A mutation is a change in the nucleotide sequence of a cell's DNA. There are two main types of mutations: nucleotide substitution and nucleotide insertion/deletion.

A nucleotide substitution is when one nucleotide base (and its complement) is replaced with a different nucleotide (and its complement). There are three types of mutations that result from a nucleotide substitution: silent, missense, and nonsense. Due to the redundancy of the genetic code, some mutations may not have any effect on the protein sequence. For example, if the codon changes from GAA to GAG, this would not make any difference, as both of these codons code for the same amino acid - glutamic acid. This is a **silent** mutation, as it does not affect the polypeptide chain. A **missense** mutation is when a nucleotide substitution changes the amino acid that is coded for. For example, if a mutation causes the second nucleotide in the mRNA codon UCU to change from C to A (now, UAU), that would change the amino acid from serine (Ser) to tyrosine (Tyr). Missense mutations can have a big impact on the function of the protein, but some are relatively harmless. A **nonsense** mutation is when an amino acid codon becomes a stop codon [24]. For example, if the second codon in UCA (which codes for serine) changes from a C to an A, UCA would become UAA (a stop codon). This would stop the protein development too soon.

Insertion and deletions are mutations in which one or more nucleotides are either inserted or deleted from the sequence. These mutations are collectively called **frameshift mutations**. Frameshift mutations can cause considerable damage due to the way in which mRNA is read and translated. Since mRNA is read in codons, which are a series of three nucleotide bases, insertions and deletions would change every single codon after the mutation, therefore changing the entire amino acid sequence. The nucleotide substitutions only affect one codon, whereas frameshift mutations would cause the entire mRNA sequence after the mutation to be regrouped into different codons [24].

Insertions, deletions, and substitutions are all **point mutations**, or single nucleotide polymorphisms (SNPs), meaning that they only involve one base (but can affect any number of amino acids). There can also be **small-scale mutations** (mutations that involve multiple bases) and **large-scale mutations** (mutations that involve larger regions). Large-scale mutations are usually seen with **segmental duplications** (large chromosomal region copied many times) and **transposable element insertions** (segment moves to a different region in the genome). Aneuploidies are another type of mutation; they are the insertion or deletion of entire chromosomes [25]. This results in extra or missing chromosomes, which can cause significant issues in the organism. The two main types of aneuploidies are **monosomy** (organism has one copy of a chromosome that they should have two copies of) and **trisomy** (organism has a third copy of a chromosome that they should only have two copies of) [26].

Mutations are caused by various factors, including mistakes in DNA replication and recombination, high-energy radiation, and chemical mutagens. Genetic variation stems from mutations through natural selection. In some cases, a mutation may lead to a protein that is beneficial to the organism's survival, so that mutation gets passed on more frequently from the parent to the offspring (due to higher survival rates). On the other hand, mutations can be harmful to the organism, causing cancer or genetic disorders. Overall, mutations drive genetic diversity and allow for species and organisms to adapt and change over long periods of time to better suit their environment.

## 3.7 Genomics

### 3.7.1 The Genome

The genome is the entire DNA sequence for an organism, encompassing all of the organism's genetic information. The human genome was first estimated to have about 100,000 protein-coding genes, but with advances in sequencing technology, it is now estimated that there are 20,000-25,000 protein-coding genes. Each one of these genes codes for three proteins, on average [27].

Genomics is the genetic mapping and sequencing of genomes. With recent advancements in technology, sequencing data can be collected without prior knowledge of the genome, allowing for new genes to be discovered.

### 3.7.2 Human Genome Project

The Human Genome Project is arguably one of the most important accomplishments in scientific history. Led at the National Institutes of Health (NIH), the international endeavor began on October 1, 1990, and ended in April 2003. Over those 13 years, the Human Genome Project sequenced the entire human genome using about 100 blood samples from the representative population. They found that there were about 20,500 human genes and were able to provide detailed information on the structure, function, and organization of these genes. Scientists had originally believed that there were anywhere from 50,000 to 140,000 genes in the human genome, so this value of 20,500 was surprising to many [28]. The partial genomic data was published in February 2001.

> "It's a history book - a narrative of the journey of our species through time. It's a shop manual, with an incredibly detailed blueprint for building every human cell. And it's a transformative textbook of medicine, with insights that will give health care providers immense new powers to treat, prevent, and cure disease."
>
> -Francis Collins, Director of the National Human Genome Research Institute

The finished sequence covers approximately 99% of the human genome and has an accuracy of 99.99%. Additionally, scientists identified about 3 million human genetic variations. Alongside the human genome, scientists sequenced the entire mouse genome sequence and published it in 2002 [30]. This data allowed scientists to better understand the roles of genetic factors in complex diseases, including cancer and diabetes, and improve therapeutic strategies and diagnostics.

# Chapter 4
# RNA Sequencing and Differential Gene Expression

## 4.1 Introduction

RNA sequencing, or RNA-seq, is a technique that uses next-generation sequencing to reveal patterns and quantify certain elements of the transcriptome. This information sheds light on the details of the genome by understanding the structure of the genes, finding novel splicing patterns, and comparing gene expression levels under different environmental or treatment conditions. A similar tool for gene expression is the hybridization-based DNA microarray. However, a microarray only surveys predefined areas of the transcriptome whereas RNA-seq allows for the sequencing of the entire transcriptome. Prior knowledge of the transcriptome is not required for

RNA-seq. RNA-seq is also more sensitive for the detection of genes with lower expression [1].

## 4.2 Experimental Design

When initially planning the experiment, many factors must be taken into consideration, such as avoiding bias and capturing enough variability. The variability can be maintained with technical and biological replicates. Technical replicates are the repeated measurement of the same sample. Biological replicates are the measurement of different samples to understand the natural biological variation. RNA-seq experiments generally use three biological replicates per sample, but this can increase based on the goal of the experiment. If differential gene expression is being measured, the number of replicates can be as high as twelve. Avoiding bias is also crucial to a well-executed experiment. This can be done with a variety of techniques, including randomization and blocking. Prior to starting the experiment, the possible sources of variability, or nuisance factors, should be identified and accounted for. Randomization ensures that unconscious bias is eliminated by relying on random chance to make certain decisions, including which cells to treat and which to use as a control. Blocking is a method used when the explanation for the differential gene expression can be inferred (such as cell cycle status or weight). The samples would be grouped into "blocks" based on the predetermined trait to increase the sensitivity of the statistical tests. In each group, the possible sources of variability would be kept constant, increasing the likelihood that the identified change in expression is a result of the condition being studied. In general, you should block all that is possible and randomize the rest [1].

## 4.3 Sequencing Library

### 4.3.1 Library Preparation – RNA Extraction

Prior to sequencing, RNA must be extracted from its environment. One method of extraction is silica-gel based membranes with ethanol. This works because only the RNA binds to the membrane and the other components are washed away. Another method of extraction is liquid-liquid extraction with phenol-chloroform. With this method, the components of the cell are dissolved into three phases: the organic phase, interphase, and aqueous phase. The RNA remains in the aqueous phase. Following the phenol-chloroform extraction is an alcohol precipitation procedure to remove the salt from the RNA and concentrate the molecules. Regardless of the technique used, RNA extraction must occur in a controlled manner, since it is a highly variable procedure [1, 2].

Library preparation is the term used to encapsulate the entire process of getting the DNA fragments ready for sequencing. The Illumina protocol will be followed for this guide. For the Illumina machine to read the DNA fragments, the cDNA must be single-stranded and in a certain range of sizes. Generally, this cDNA range is between 150 to 300 base pairs (bp). A library preparation protocol must be chosen based on the needs of the experiment. For example, the details of smaller transcripts may be lost. If smaller transcripts are an area of focus in the experiment, a different protocol should be selected [2].

Illumina estimates that library preparation takes six hours in total with three hours "hands-on" time. The four main steps outlined are as follows: fragmenting the DNA, repairing its ends, ligating adapters, and selecting ligated DNA. It begins with the isolated RNA from the tissue sample. The RNA is broken into small fragments, double-stranded cDNA is synthesized from these fragments, and the cDNA is amplified by the polymerase chain reaction (PCR). The sequencing adapters are added onto the strands of cDNA [3].

### 4.3.2 Cluster Generation

After adapters are added onto the cDNA strands, the second phase may begin: cluster generation. The Illumina protocol estimates that cluster generation takes about four hours with less than ten minutes of "hands-on" time when testing up to 96 samples. The adapters on the cDNA guide the way in which they attach to the flow cell. The patterned flow cell is the glass slide that has small chambers. The advantages of this new technology are that the small wells allow for higher data output, more reads, and a faster procedure overall. After positioning the flow cell, the adapters complementary to the ones attached to the cDNA are deposited into the cells. The adapters on the cDNA bind to their complements on the flow cell. After the cDNA attaches by complementary adapters, it bends over so the adapter on its other end can attach to *its* complementary adapter on the flow cell. A primer makes a copy of the strand (according to classic base-pairing rules), creating a forward strand and a reverse strand. The bridge amplification process generates clusters. This occurs very rapidly in order to prevent other cDNAs from binding, which would result in a polyclonal cluster [3]. Amplification in this manner ensures that there is a monoclonal cluster in each well. This is important because it means that each well has clusters that originate from a single template. Once the bridge amplification and cluster generation are complete, the samples are ready for sequencing.

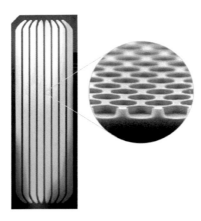

Figure 4.1 **Illumina flow cell**. From https://www.illumina.com/

### 4.3.3 Sequencing

The sequencing step is exactly what the name entails: using the Illumina sequencer to sequence the samples! There are two ways to sequence the samples: single-read and paired-end. The main difference between these two is the depth of the sequencing. In single-read sequencing, the sequencer only reads the fragment from one end to another. In paired-end sequencing, the sequencer reads the fragment from one end to another and then repeats this process but beginning at the opposite end of the fragment. Paired-end sequencing has many advantages, including higher quality, longer reads, and the detection of novel transcripts, gene fusions, and genomic rearrangements. Longer reads are favorable since they provide better information on the location of base pairs. With this increased quality of detection comes a higher price, as paired-end sequencing is generally more expensive than single-read sequencing [3].

The depth of coverage is another important parameter to consider. It represents the number of times that a site on the genome has been sequenced during a single run in relation to the genome size. The Lander-Waterman equation is commonly used for this calculation [1].

$$depth\ of\ coverage = \frac{read\ length \cdot number\ of\ reads}{haploid\ genome\ length}$$

The quality of the data and the number of times that the base has been sequenced (coverage) are directly related, meaning that increased coverage results in increased quality. In RNA-sequencing experiments, it is recommended that the coverage be at least 20 million reads per sample, with some experiments requiring up to 100 million reads [4]. This number is based on the least abundant RNA species of interest to the experiment, but it is unknown at the beginning of sequencing. Therefore, estimates must be made based on the transcriptome size, error rate for the sequencer, type of experiment/question, and scientific literature. If the goal is to identify genes with low expression, small fold changes between different samples, and detect novel and chimeric transcripts, it is recommended to have higher coverage [1].

## 4.4 Raw Data

### 4.4.1 Output

The output of the Illumina sequencer will be two **FASTQ** files per sample in a paired-end run (one FASTQ file per sample in a single-read run). The two FASTQ files are written as R1 and R2, respective to the direction of sequencing. The data is presented in four lines for each strand. The first line is a sequence identifier with information about the specific cluster. The second line is the sequence itself using the bases A, C, T, G, and N. The third line is simply one plus (+) sign. The fourth line is the quality scores for the base calls encoded with ASCII characters [5].

### 4.4.2 Quality Control

The overall quality of the sequenced reads should be checked before any further analysis is conducted. Low-quality reads may have certain indicators, including PCR duplicates, unmappable reads, contamination, and unwanted reads from rRNA/tRNA. The program FASTQC can detect all of these issues with the exception of unmappable reads.

It analyzes the quality scores of each read and gives each a quality rating.

## 4.5 Aligning Reads to the Reference Genome

The position and origin of the cDNA fragments must be established in order to contextualize the data. In simpler terms, the origin of the read must be identified. This process is called read alignment, or mapping, and it is done before any exploratory analysis can take place. It involves matching the short sequence of nucleotides from the raw data to the long sequence of nucleotides on the reference genome. The main challenge is time and accuracy, including solving gaps and mismatches in the sequence, but many aligners are competing to optimize both of these factors. The speed and accuracy of the aligner also depend on the quality of the raw data, which changes based on the sequencer. Generally, Illumina has a lower error rate than its counterparts, such as Nanopore (long-read data) [1].

Specific to RNA-seq data, the presence of "multiple different isoforms" of the same gene presents an issue with some alignment programs. Isoforms are proteins that are very similar in their amino acid sequence and are encoded by the RNA transcript of the same gene, but with different exons removed. There is also the issue of mapping ambiguity, in which the reads overlap with more than one part of the transcriptome. This problem is mitigated by identifying novel spruce events and noting the location of known introns [1].

There are many RNA-seq alignment programs, including STAR, HISAT2, Bowtie2, and TopHat. STAR generally has the highest sensitivity/accuracy and shortest run time when compared with other popular aligners. HISAT2 is also fast and accurate, but it requires more memory than STAR. TopHat2 is slower than both the aforementioned aligners, but it requires much less memory [1]. There are advantages and drawbacks to each aligner, so the one most appropriate for the context of the experiment and limitations of the technology must be chosen.

### 4.5.1 Aligning with STAR

The index files for the reference genome contain all of the information in a compressed format, including the names/lengths of chromosomes, gene information, and splice junctions. The reference genome works alongside an annotation file.

The parameters for the genome indices are as follows:

| Options/Parameters | Purpose |
|---|---|
| --runThreadN | Specify the number of threads |
| --runMode | The mode used to generate genome |
| --genomeFastaFiles | The path to the FASTA files |
| --genomeDir | The path to store genome indices |
| --sjdbGTFfile | Annotation file |
| --sjdbOverhang | Read length minus 1 |

The alignment matches the reads to the reference genome by using the information in the annotation files. In paired-end sequencing, each sample generates two FASTQ files. STAR can merge both R1 and R2 files if both file names are specified in the original commands.

STAR searches for the longest sequence on the read that matches the reference genome. These sequences are called the Maximal Mappable Prefixes (MMPs). Seeds are the sections of the reads that are broken apart to be mapped to the genome separately. The first seed (first MMP) would be mapped to the reference genome. STAR would then search the reference genome for a match to the unmapped portion of the original read. STAR does this by using an uncompressed suffix array. STAR then has to put the seeds together, and this process is called stitching. This is done using the relative scores of each read [6].

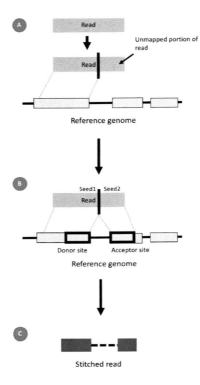

Figure 4.2. **STAR Aligner.** STAR (Spliced Transcripts Alignment to a Reference) is an aligner used to map reads to a reference genome. There are two distinct steps in the process. (A) Seed searching is when STAR searches for the longest sequence (Maximal Mappable Prefixes – MMPs) that matches a location on the reference genome. The first MMP is called seed1. (B) STAR then searches for the unmapped part of the read to match to the reference genome. This is called seed2. (C) Clustering, stitching, and scoring are the next steps. The proximity of the seeds to the anchor seeds are measured, and the seeds are stitched together based on their realtive scores for mismatches and gaps [6].

## 4.5.2 SAM/BAM Files

After the alignment, the output should be SAM/BAM files. The Sequence Alignment Map (SAM) file format is the human-readable alignment form while Binary Alignment Map (BAM) files are their binary equivalent. BAM files are compressed and easier to work with, as the format allows for BAM files to be indexed in a way that the reads can be retrieved without loading the entire file. SAM files have an optional header section with basic information about the file and an alignment section.

The SAM file header section has lines beginning with an "@" character and an abbreviation. The "@SQ" line would have information on the reference sequences. The record types for "@SQ" include SN (reference sequence name), LN (reference sequence length), AN (alternative reference sequence names), AS (genome assembly identifier), SP (species), and DS (description). Other sections include "@HD," which is the header line specifying the sorting order and the format version, and "@PG," which describes the program name/version [7].

In the alignment section, 11 fields appear in the following order [7]:

<QNAME> <FLAG> <RNAME> <POS> <MAPQ> <CIGAR> <MRNM> <MPOS> <ISIZE> <SEQ> <QUAL>

- QNAME: read name
- FLAG: describes alignment
    - 1: mapped
    - 2: mapped as part of pair
    - 4: unmapped

- o 8: mate is unmapped
- o 16: mapped to reverse strand
- o 32: mate reverse strand
- o 64: first in pair
- o 128: second in pair
- RNAME: reference sequence name
- POS: 1-based leftmost mapping position of alignment
- MAPQ: alignment quality
- CIGAR: numbers and letters describing edits/operations used in alignment. The numbers indicate the base length for the operations and the letters indicate the operations.
  - o M: Mismatch
  - o I: Insertion
  - o D: Deletion
  - o N: Skipped
  - o S: Soft Clipped
  - o H: Hard Clipped
  - o P: Padding
- MRNM: reference name of the next alignment
- MPOS: leftmost position of next alignment
- ISIZE: inferred insert size of alignment
- SEQ: raw sequence
- QUAL: quality scores for each base in the sequence

## 4.5.3 Integrative Genomics Viewer (IGV)

The Integrative Genomics Viewer (IGV) is a tool from the Broad Institute of MIT and Harvard for the visualization for large genomic datasets, including

next-generation sequencing data. It facilitates quality assessments for the alignment. You can learn more and download IGV using this link:
http://software.broadinstitute.org/software/igv/

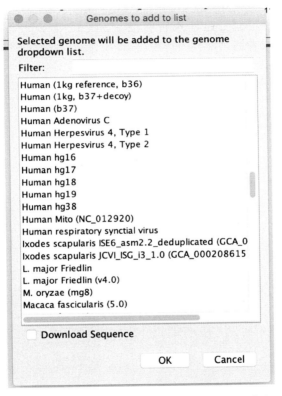

Figure 4.3. IGV has many genomes already hosted on their server. If the genome that you need is not already loaded into IGV, you can use your own FASTA file for the reference genome.

## 4.6 Read Quantification

The genomic coordinates of the mapped reads are located in the BAM file, which must be coupled with the genomic coordinates for the reference (GTF file). One of the most accurate and easy-to-use tools is featureCounts. It counts the reads that are mapped to a single location and assigns the reads to a gene/exon to output the gene counts.

There are multiple parameters required when running featureCounts.

| | |
|---|---|
| -T 4 | # specifying four cores (value based on number of cores set in the beginning) |
| -s 2 | # value represents if data is stranded or not ("2" : reverse-stranded) |
| -a | # include path to GTF file |
| -o | # include path to output |
| \~ | # list of all BAM files for the count data |

The output of featureCounts is two files: a count matrix and a summary file.

| GENE | CTRL-1 | CTRL-2 | CTRL-3 | CTRL-4 | TREAT-1 | TREAT-2 | TREAT-3 | TREAT-4 |
|---|---|---|---|---|---|---|---|---|
| ENSMUSG00000085990 | 10 | 12 | 11 | 6 | 9 | 20 | 13 | 8 |
| ENSMUSG00000023367 | 7 | 6 | 4 | 10 | 6 | 4 | 5 | 10 |
| ENSMUSG00000034871 | 3 | 8 | 2 | 1 | 5 | 5 | 3 | 6 |
| ENSMUSG00000001313 | 2 | 2 | 7 | 5 | 3 | 1 | 2 | 3 |
| ENSMUSG00000049295 | 15 | 10 | 8 | 4 | 8 | 5 | 5 | 9 |
| ENSMUSG00000107802 | 6 | 4 | 10 | 5 | 13 | 10 | 10 | 19 |
| ENSMUSG00000087528 | 5 | 3 | 7 | 3 | 9 | 7 | 3 | 6 |
| ENSMUSG00000044066 | 66 | 71 | 71 | 53 | 69 | 79 | 62 | 76 |
| ENSMUSG00000032567 | 1 | 1 | 1 | 1 | 3 | 2 | 1 | 1 |

Table 4.1. **Gene count for each sample.** This table includes the read counts of the first nine genes for my personal research project. There were eight samples: four control and four treated. Each gene's count values are shown for each sample. Each row is a gene and each column is a sample. The genes in the gene column are represented by their ENSEMBL gene IDs.

## 4.7 Normalizing Read Counts

The read counts from featureCounts need to be normalized before any analysis can be done. This is mainly due to the differences in library size and library composition.

There are differences in library sizes between samples, which must be adjusted for. These library sizes are the total number of reads per sample for all of the genes in that sample. These values should be the same for the differences in each gene's read counts to be due to biological differences and not differences in sequencing depth [8].

Differences in library composition must be accounted for, as RNA-seq experiments can compare one tissue type to another. Genes that are specific to one tissue would show higher counts when compared to genes not specific to that tissue, and vice versa. This is an issue when these tissue-specific genes are expressed in large quantities, as they dominate the sample's read counts and make a comparison between two samples more difficult [8].

DESeq2 and edgeR are two popular tools that are used to normalize the read counts by accounting for the two aforementioned problems.

### 4.7.1 DESeq2

1. Calculate the natural log of every read count.

2. Take the average of each row (rows = genes). Logs are used so that outliers do not sway the average as much (when compared to averaging raw read counts). The averages of log values are called geometric means.

3. Filter out genes that have infinity as their geometric mean. This is the result of a gene having "zero" as one of their read counts. This mitigates the issue of differences in library composition and allows the scaling factors to be more representative of the genes of interest.

4. Subtract the log of the read counts by the average log value for each gene. Basic log rules show that $\log(x) - \log(y) = \log\left(\frac{x}{y}\right)$. Therefore, the difference between the log values would actually be the ratio between the reads in each sample and the average across all samples for the same gene.

5. Find the median of these log-ratios for each sample. Using the median, as opposed to the mean, helps mitigate the effect of outliers. The

median should be taken per column (per sample).

6. Find the scaling factor for each sample. Convert the log medians back by raising e to the power of the median. For example, if the median of a sample is 0.4, then the scaling factor would be $e^{0.4}$, which is approximately 1.492.

7. Divide the original read counts by the scaling factor for each sample (scaling up vs. scaling down) [9, 10].

In statistical tests, a p-value of less than 0.05 is considered significant. For example, a p-value of 0.05 means that there will be a false positive 5% of the time. When testing entire genomes for differential gene expression, these false positives could add up very quickly. Five percent of 20,000 genes is 1,000 potential false positives, meaning 1,000 genes are being falsely reported as differentially expressed. This is an issue known as the multiple test problem. To mitigate the effects of this problem, the False Discovery Rate (FDR) and the Benjamini-Hochberg method can be used. DESeq2 incorporates both of these methods to mitigate the multiple test problem.

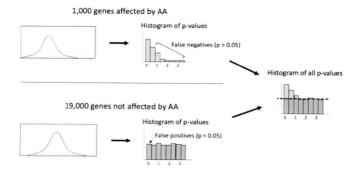

To demonstrate the false discovery rate, an example of 20,000 genes can be used in which 1,000 are known to be affected by a treatment and 19,000 are not. The histogram of the p-values for the genes affected by the treatment is shown in the first histogram. Most of these p-values, as expected, would be less than 0.05. However, the point at which both distribution curved overlap would produce some false negatives. Ideally, all of the p-values would be less than 0.05, but this is not the case. In the 19,000 genes not affected by the treatment, the p-values would be distributed evenly across all intervals, but this causes some to be less than 0.05. The genes in this group are false positives since they would be falsely reported as statistically significant ($<0.05$). The histogram of the p-values of all 20,000 genes is the sum of the two previous histograms, and here, the real positives are shown above the dotted line. To distinguish these real positives from the false positives, the Benjamini-Hochberg method is used [11].

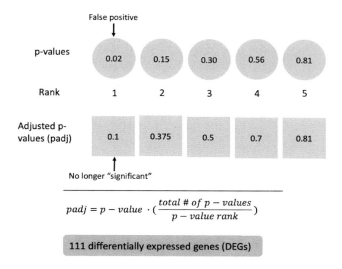

$$padj = p - value \cdot \left(\frac{total \; \# \; of \; p - values}{p - value \; rank}\right)$$

111 differentially expressed genes (DEGs)

The Benjamini Hochberg method is essentially an algorithm to interpret the original p-values and create adjusted p-values to assess the true statistical significance. The method has multiple steps, and the parameters can be as strict or as lenient as the experiment requires. In the example above, there are five p-values, with one being less than 0.05 even though it originates from the previous group of 19,000 genes that are not affected by the treatment. This is a false positive.

1. The p-values are ranked from smallest to largest, with the smallest p-value being ranked as 1.

2. The adjusted p-values are calculated based on their relative ranks and original p-value. The equation is shown above, as the adjusted p-value is generally equal to the total number of

p-values divided by the p-value's rank multiplied by the original p-value.

    a. The largest p-value (highest ranked) is the same as its adjusted p-value.

    b. The adjusted p-value for the next highest-ranked is either the previous adjusted p-value or the product of the aforementioned equation. Whichever value is smaller becomes the adjusted p-value. This continues from the highest-ranked p-value to number 1.

3. Set parameters to identify the true positives in this list (ex. padj < 0.05) [10, 11]

In the example above, the original p-value of 0.02 (false positive) now has an adjusted p-value of 0.1, which is no longer considered statistically significant. The rest of the p-values remain above the threshold and are still not being reported as significant.

## 4.7.2 edgeR

Like DESeq2, edgeR also accounts for the differences in library composition and sequencing depth between samples, but with different methods of normalization.

1. Remove all genes that have 0 as their read count values (this means they were not transcribed).

2. Identify a reference sample used to normalize all other samples. (1) First, each sample should be scaled by its read counts by adding up the read counts for each sample and dividing each read count by this sum. (2) In each sample, find the value that is greater than or equal to 75% of the values in that sample (75th quantile). (3) Take the average of the 75 quantiles found in part b (there should be one number per sample). The reference sample is the sample who's 75th quantile is closest to this average.

3. Identify the genes to calculate the scaling factors. Each sample will do this relative to the reference sample. This means that the selected genes will not be the same in all of the samples. edgeR has to remove extreme and biased genes from the set. The method for doing so is taking the log fold changes of the scaled reads and setting an appropriate threshold. The equation for assessing these log-fold changes is $log2(\frac{scaled\ read\ on\ reference\ for\ gene\ x}{scaled\ read\ on\ sample\ for\ gene\ x})$. If the value is infinity, then the gene is removed, as this means that there are no reads mapped to these genes in either sample.

4. Identify the genes that are transcribed either highly or lowly in both of the samples. Calculate the geometric mean for each gene by taking the average of the log2 for the scaled read counts in each sample. Remove genes with infinity values for their geometric means.

5. The two data sets now are the list of potentially biased genes and the list of the geometric means of the log2 values. These lists can be filtered out according to the needs of the experiment, but some suggest filtering out the top and bottom 30% of the list of the biased genes and the top and bottom 5% of the highly/lowly transcribed genes list. The scaling factor will be calculated based on the genes that remain in *both* lists.

6. Calculate the weighted mean of the remaining log2 ratios (for each sample).

7. Raise 2 to the power of the weighted average of the log2 ratios. This converts the value from a log to a scaling factor. This scaling factor is specific to each sample.

8. Center/shift the scaling factors around 1 by dividing the original scaling factors by their geometric mean [12, 13].

Counts per million (CPM) is important to note, as edgeR recommends removing any genes with a CPM value of one or fewer in at least two samples. CPM is calculated by dividing the reads for a specific gene in one sample by the total number of reads in that sample (divided by one million). For example, if gene X had 16 reads in one sample, and that sample had 3,589,000 reads total, then the expression would be $\frac{16}{3.589}$, which is equal to 4.458. Since this CPM is greater than 1, it would not be removed (given that it has a CPM greater than 1 in another sample). Sequencing depth is important for CPM, as the denominator is based on the number of total reads for that sample. CPM cutoffs can be adjusted based on sequencing depth, as the CPM scaling factor can either be too big or too small depending on the total number of reads. For example, consider a gene that has five reads in a sample that has 10 million total reads: the CPM scaling factor is 10 ($\frac{10,000,000}{1,000,000}$) and the CPM is 0.5 ($\frac{5}{10}$). Another example is a gene that has five reads in a sample that has 40,000 total reads: the CPM scaling factor is 0.04 ($\frac{40,000}{1,000,000}$) and the CPM is 125 ($\frac{5}{0.04}$). Different CPM cutoffs should be tested after the p-values have been calculated to determine the most appropriate cutoff for the data [1].

The main difference between edgeR and DESeq2 is their methods of determining which genes make the cutoff and which don't. DESeq2 examines the average reads across all of the samples and keeps the gene if the average of their read counts is above the cutoff. On the other hand, edgeR requires that at least two samples be above the cutoff to keep the gene. This means that edgeR can detect outliers better than DESeq2 can, but DESeq2 uses a different procedure to detect outliers and does not rely on this filtering step. DESeq2 uses quantiles instead of the minimum CPM threshold when comparing it to the number of significant genes. This is advantageous because CPM depends on sequencing depth, while quantiles do not (since quantiles are based on percentages) [1].

## 4.8 Exploratory Data Analysis

### 4.8.1 Principal Component Analysis (PCA)

Principal Component Analysis (PCA) is a dimension reduction technique used to understand and visualize high-dimensional data.

Measurements of one gene can be plotted on a one-dimensional number line, with each sample representing one dot.

Measurements of two genes can be plotted on a two-dimensional graph with the x-axis representing one gene and the y-axis representing the other gene.

Measurements of three genes can be plotted on a three-dimensional graph with the x-axis representing one gene, the y-axis representing the other, and the z-axis representing the third gene. This is where it starts to get complicated.

After the third gene, you hit a wall; you can no longer represent the data for these genes on a plot that makes sense to the viewer. You can't keep adding axes with every gene, and there are approximately 20,000 genes in the human genome.

For this reason, the data must undergo dimension reduction. In doing so, we are hoping to reduce the noise while extracting the key features, or *principal components*, of the data while preserving its variability. The variability refers to how spread out the data is. PCA takes the normalized, high-dimensional data and reduces the data from the original ~20,000 genes to just two axes (principal component 1 and principal component 2).

To understand how PCA extracts these two principal components, we can use an example of just two genes. As we mentioned before, you can plot one gene on the x-axis, and the other gene on the y-axis. The average measurement for gene 1 (x-axis) and gene 2 (y-axis) can be calculated by simply averaging the read counts for each gene. That new coordinate (x,y) is the center of the data. Then, you can shift the data so that the coordinate is now on the origin (0,0). With the data centered on the origin, you can fit a line to the points. PCA starts with a random line going through the origin and continues to rotate the line until it fits the data best. The mechanism through which PCA decides on this best fit is based on simple mathematical principles. PCA projects the data onto that first random line and measures the distances from the points to that line. The best fit line would be the one that maximizes the sum of the squared distances from the projected points (on the line) to the origin.

The Pythagorean theorem states $a^2 = b^2 + c^2$. The following definitions will be used in the explanation:

- a = distance from origin to original point
- b = distance from point to line
- c = distance from origin to projected point (on the line)

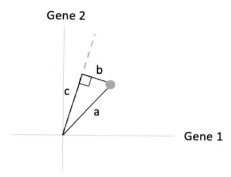

The value of $a^2$ will be the same regardless of the line being used. The value of $b^2$ will decrease as the line gets closer to the point, and the value of $c^2$ will increase as the line gets closer to the point. The entire goal is to make the line as close to the points as possible, and there are two options to do so: minimize $b^2$ or maximize $c^2$. PCA uses the latter option and finds the best fit line by maximizing the sum of the $c^2$ values for every point (we use $c^2$ instead of c so that negative values are not a part of the calculation). The sum of $c^2$ for the points is called the Eigenvalue for the principal component.

The line that best fits all of these data points is called principal component 1 (PC1). A large slope for PC1 ($m > 1$) indicates that gene 2 accounts for most of the variability (as data is more spread out along the y-axis). A small slope for PC1 ($m < 1$) indicates that gene 1 accounts for more of the variability (as data is more spread out along the x-axis). This slope ratio is essentially the linear combination of the two genes. The distances for the principal components need to be scaled up/down to 1 by dividing all three sides of the triangle by the distance of PC1. The ratio for the slope remains the same. This new vector is called the Eigenvector for PC1 and the elements of the ratio are called the loading scores. As mentioned before, the sums of squares used to determine the best file line is called the eigenvalue for PC1 and the square root of this eigenvalue is called the singular value for PC1.

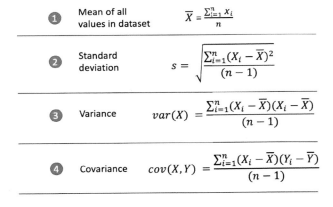

Figure 4.4 Statistical foundation of principal component analysis.

Principal Component 2 (PC2) is the line perpendicular to PC1 that passes through the origin. This means that the slope for PC2 is the opposite and reciprocal (or "op-reciprocal," as my freshman year math teacher said) of the slope for PC1.

PC1 and PC2 become the new axes of the plot (the original plot gets rotated to accommodate for this). The projected x and y values of the original samples get plotted on the graph.

As mentioned in the introduction to this visualization technique, PCA attempts to preserve variability in the data in each of its principal components. To calculate this variation for each PC, the eigenvalues are divided by the (sample size minus 1). A scree plot is a bar graph of the percent variance that each principal component accounts for.

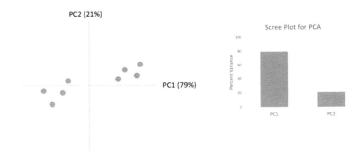

Figure 4.5 Example Scree Plot for PCA with corresponding points.

For three genes, you would find PC1, PC2, and PC3 by finding the perpendicular lines to the previous PC and calculating the best fit lines. You then use the eigenvalues to determine the percent variance that each PC accounts for. You would take the two principal components that account for the most and the second most variance and construct the PCA plot with these two principal components as the x- and y-axis. In this example, let us imagine that PC1 accounts for 68% of the variance, PC2 accounts for 22% of the variance, and PC3 accounts for 10% of the variance. The PCA plot would be constructed using PC1 and PC2, as these two PCs account for the most variance together (90%). This applies to PCA plots for 4 genes, 5 genes, 10 genes, or 20,000 genes [14].

PCA can be performed with the function prcomp().

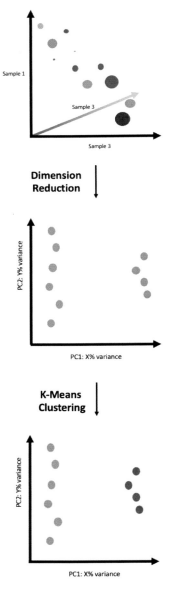

Figure 4.6 PCA and K-means clustering.

PCA plots are used to identify clusters of data. Generally, K-means clustering is used for this analysis. The first step is to identify the number of clusters you want in your data, which becomes the K value. A K value of 2 would mean that you want to identify two clusters. Then, two data points are selected, as these become the temporary clusters. The distances from each point to the two clusters are measured. The points are assigned to whichever cluster is closest to the point. The mean of each cluster is calculated, and the measurement/clustering process is repeated for these means. The variation within the clusters is calculated and recorded. This entire process is repeated with different initial clusters (pick two points to be clusters, measure distance from points to clusters, calculate mean, cluster based on means, and calculate variance). The final clusters would be chosen based on which trial had the least variance.

The K value is chosen through trial and error. As the K values increase, the variation decreases until K equals the total number of points (which would result in a variation of zero because there would be one point per cluster). Each K value (K=1, K=2, K=3, etc.) is assessed relative to its variation, and the variations/distances are plotted on an elbow plot (with the value of K as the x-axis) to find the ideal K value, which would be the point at which the variation undergoes the most reduction [15].

The function for K-means clustering in R is kmeans(). Its default is to only try one set of clusters, so the parameter nclust has to be set accordingly. The number of starting points is specified by nstart.

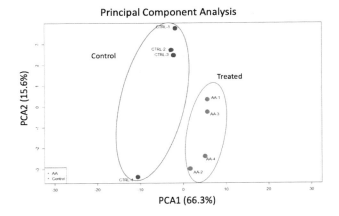

Figure 4.7 **Example PCA plot for RNA-seq data.** Two clusters are identified, one for the control samples (in red) and one for the treated samples (in blue).

## 4.9 Differential Gene Expression Analysis

The PCA plot visualizes the difference in gene expression between the clusters (representing samples), but we also need to know the magnitude and significance of the difference. Linear regression models applied to each gene can be used for DGE. The form for this model is usually $Y = b_0 + b_1 \cdot x + e$, with Y equal to the total read counts for the gene, b0 representing the intercept, b1 equal to the coefficient for the difference between samples, x representing the condition (0 or 1), and e referring to the error [1].

DESeq2 and edgeR use a negative binomial model, rather than a linear model, to determine the differences in gene expression. A more specific method is using the Poisson distribution. With a Poisson distribution, the mean would be equal to the variance. This characteristic is only convenient for technical replicates, not biological replicates, as biological replicates have more variance that the Poisson distribution does not capture. The negative binomial distribution is better equipped to handle this greater variance; the variance can be greater than the mean when applicable [1].

| Gene | baseMean | log2FoldChange | lfcSE | stat | pvalue | padj | Ctrl-1 | Ctrl-2 | Ctrl-3 | Ctrl-4 | AA-1 | AA-2 | AA-3 | AA-4 |
|---|---|---|---|---|---|---|---|---|---|---|---|---|---|---|
| ENSMUSG00000085990 | 776.97 | 0.584 | 0.094 | 6.20 | 6E-10 | 1E-07 | 681.68 | 592.45 | 622.97 | 589.70 | 1016.05 | 863.26 | 942.17 | 907.48 |
| ENSMUSG00000023367 | 133.53 | -0.587 | 0.155 | -3.78 | 2E-04 | 5E-03 | 171.42 | 142.22 | 142.98 | 184.92 | 124.68 | 91.39 | 110.42 | 100.22 |
| ENSMUSG00000034871 | 280.50 | 0.588 | 0.110 | 5.34 | 9E-08 | 9E-06 | 221.25 | 216.63 | 242.04 | 216.49 | 337.75 | 330.28 | 346.73 | 332.84 |
| ENSMUSG00000001313 | 484.86 | 0.591 | 0.111 | 5.30 | 1E-07 | 1E-05 | 420.57 | 342.85 | 365.61 | 419.44 | 511.75 | 625.28 | 556.22 | 637.17 |
| ENSMUSG00000049295 | 143.13 | 0.594 | 0.161 | 3.68 | 2E-04 | 7E-03 | 99.66 | 95.13 | 136.85 | 125.16 | 203.77 | 141.16 | 180.59 | 162.74 |
| ENSMUSG00000107802 | 234.33 | 0.595 | 0.152 | 3.90 | 1E-04 | 3E-03 | 189.35 | 206.27 | 221.61 | 128.54 | 316.35 | 232.56 | 276.56 | 303.41 |
| ENSMUSG00000087528 | 579.66 | 0.608 | 0.103 | 5.92 | 3E-09 | 6E-07 | 475.38 | 457.76 | 508.59 | 394.64 | 713.66 | 638.85 | 719.27 | 729.11 |
| ENSMUSG00000044066 | 163.55 | 0.612 | 0.141 | 4.35 | 1E-05 | 7E-04 | 136.53 | 153.53 | 121.53 | 104.86 | 197.26 | 196.36 | 196.07 | 202.28 |
| ENSMUSG00000032567 | 102.62 | 0.615 | 0.168 | 3.65 | 3E-04 | 7E-03 | 90.69 | 89.48 | 82.72 | 60.89 | 120.03 | 142.07 | 115.58 | 119.53 |

Table 4.2 Table generated by statistical analysis on raw reads for nine example genes.

The table generated by the statistical analysis on the raw reads shows the base mean, log2 fold change, original p-value, adjusted p-value, and normalized read counts. With this information, the genes can be filtered by setting parameters for the adjusted p-values and the log2 fold change. In order to filter the significant DEGs, the parameters can be set at p-adj < 0.05 or p-adj < 0.01 and |log2FC| > 0.58 (at least 50% change) or |log2FC| > 1 (at least 100% change or 2-fold change). The parameters can be adjusted based on the values in the dataset and which level of specificity would be most appropriate. For my own analysis on a dataset, I set p-adj < 0.01 and |log2FC| > 0.58. This reduced the original ~20,000 genes to just 110 DEGs. These were the genes that had the most significant differential gene expression between the untreated and treated samples.

## 4.9.1 Heatmaps

Heatmaps visualize the expression results for the samples in regard to their rlog-normalized read counts. To generate heatmaps, the following commands can be used:

- ggplots::heatmap.2()
  - install.packages("ggplot2")
  - library("ggplot2")
- NMF::aheatmap()
- pheatmap::pheatmap()

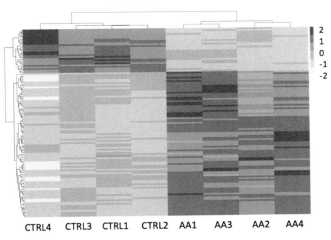

Figure 4.8 **Heatmap of all DEGs across samples with hierarchical clustering.** This heatmap and hierarchical clustering show the differentially expressed genes across all of the samples, with red being the most upregulated and yellow being the most downregulated. Each row represents a differentially expressed gene. Hierarchical clustering is represented by the groups of lines on the top and left side of the heatmap. These clusters sort the genes based on gene expression levels relative to one another.

## 4.9.2 MA Plot

An MA plot is a scatter plot with the mean of the normalized counts on the x-axis and the log fold change on the y-axis. The DEGs are highlighted in red. You can make an MA plot with the plotMA() function. The primary arguments included are as follows: object, array, coef, xlab, ylab, main, status, zero.weights() [1].

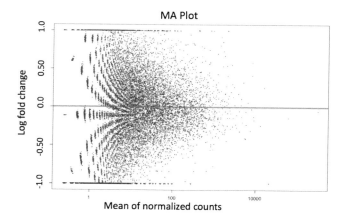

Figure 4.9 **MA-plot from base means and log-fold changes.** This is a scatter plot of log2 fold changes and the means of normalized counts. It visualizes the degree of difference between the read counts in the treatment types (control, treated).

## 4.9.3 Volcano Plot

A volcano plot is a scatterplot with the log2 fold change values on the x-axis and the adjusted p-value on the y-axis. It helps visualize the most upregulated and downregulated genes from the DEGs [1].

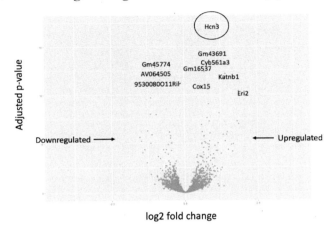

Figure 4.10 **Volcano plot showing upregulation and downregulation of DEGs.** This allows for easy identification of DEGs and visualizes the magnitude of their significance. The most upregulated genes are toward the right, and the most downregulated genes are toward the left. The top ten statistically significant genes are labeled.

### 4.9.4 Read Counts for Single Genes

Read counts for single genes can be used to assess the difference in count values between the samples for just one gene of interest.

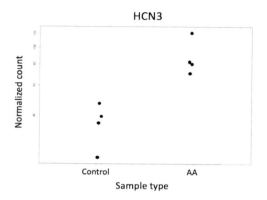

Figure 4.11 Read counts for HCN3. This plot shows the difference in normalized count values between the control and treated samples for the gene HCN3. This helps explore the difference in expression for just one gene.

## 4.10 Gene Set Enrichment and Pathway Analysis

### 4.10.1 GSEA

Gene Set Enrichment Analyses are used to find and assess the associations between differentially expressed genes and biological pathways in the body. Common pathway databases include KEGG, MSigDB, and Gene Ontology.

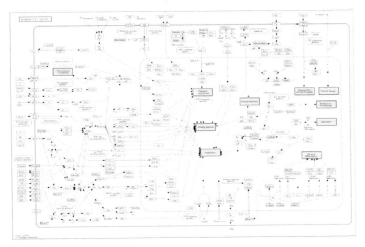

Figure 4.12 **Pathways in Cancer from KEGG Pathways.** From https://www.genome.jp/kegg-bin/show_pathway?map05200

GSEA is a tool used from UC San Diego and the Broad Institute for enrichment analysis that has both a GUI (graphic user interface) implementation and an R-based implementation. You need to register for GSEA/MSigDB to obtain the software and data sets. MSigDB refers to the Molecular Signatures Database, which is a collection of annotated gene sets to use with the software. The gene sets available belong to one of the following 8 classifications: hallmark gene sets (H), positional gene sets (C1), curated gene sets (C2), regulatory target gene sets (C3), computational gene sets (C4), GO gene sets (C5), oncogenic gene sets (C6), and immunologic gene sets (C7).

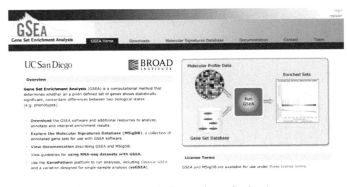

From: https://www.gsea-msigdb.org/gsea/index.jsp

## 4.10.2 PANTHER

Protein Annotation Through Evolutionary Relationship (PANTHER) is a classification system that uses data about gene function, pathways, and ontology to analyze data from gene expression and sequencing experiments. Enter your list of differentially expressed genes either as a file or in the given text box. Select the organism and operation you want to perform. The options are as follows: functional classification viewed in gene list, functional classification viewed in graphic charts, statistical overrepresentation test, and statistical enrichment test. The results will include the gene ID, mapped ID, gene name/gene symbol, PANTHER family/subfamily, PANTHER protein class, and species. PANTHER also shows its molecular function, biological processes, slim cellular component, protein class, and pathways.

PANTHER has three steps: (1) enter the gene list and specify the list type, (2) select the organism, and (3) select the type of analysis you want to conduct. From: http://www.pantherdb.org/

PANTHER finds the relevance of the entered genes in many different categories: pathway, molecular function, biological process, cellular component, and protein class. It creates pie charts to represent this data. Below are some results from an example set of 35 genes.

Figure 4.13 PANTHER Pathway results for example set.

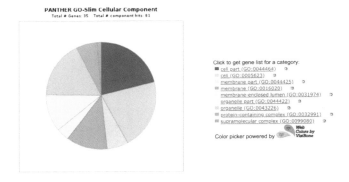

Figure 4.14 PANTHER GO-Slim Cellular Component results for example set.

## 4.10.3 Enrichr

Enrichr is a web-based enrichment tool. It has a feature called "Gene Search," as you can see in the top menu bar, that allows for you to enter the name of a gene and find information on its transcription, pathways, ontologies, diseases/drugs, cell types, legacy, and crowd.

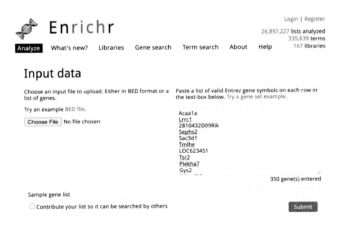

Either enter the gene list in the text box or as a file and click submit. From: https://amp.pharm.mssm.edu/Enrichr/.

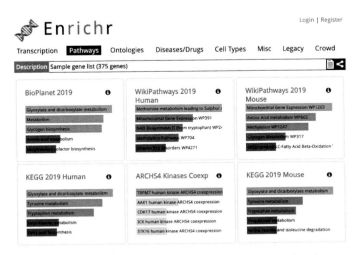

The result of pathway analysis on a sample gene list of 350 genes from Enrichr. The results show pathways from BioPlanet 2019, WikiPathways 2019 Human, KEGG 2019 Human, etc.

For example, clicking on WikiPathways 2019 Human shows a larger, more detailed view of the pathways with a table, bar graph, and clustergram that can be sorted based upon their p-values, rank, and combined score. The clustergram below shows the genes as rows and the enriched pathways as columns. The gene is associated with the pathways if the cell is colored. In the example, it shows that the gene KMO is associated with NAD Biosynthesis II (from tryptophan) WP2485 and Tryptophan catabolism leading to NAD+ Production WP4210.

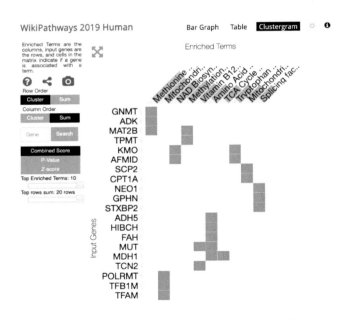

Clustergram of WikiPathways 2019 Human results for example gene list.

## 4.10.4 Fisher's Exact Test and Jaccard Index

The fundamental principles of pathway analysis are Fisher's Exact Test and the Jaccard Index. The Fisher's Exact Test assesses the significance of relationships between DEGs and members of certain pathways. The process is essentially based on a 2x2 contingency table with the options "in the pathway" and "not in the pathway" crossed against "in the gene list" and "not in the gene list" [16]. This gene list represents the differentially expressed genes based on the parameters set for the adjusted p-values and the log2 fold change. The Jaccard Index is simply a measure of similarity between two groups. The Jaccard coefficient is calculated by dividing the number of members shared by both sets by the total number of members in both sets (intersection of two sets divided by the union of two sets) [16].

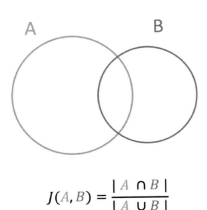

$$J(A, B) = \frac{|A \cap B|}{|A \cup B|}$$

# References

Chapter 1: Bioinformatics
1. Bayat, A. (2002). Science, medicine, and the future: Bioinformatics. *Bmj, 324*(7344), 1018-1022. doi:10.1136/bmj.324.7344.1018

Chapter 2: R Programming
1. What is R? (n.d.). Retrieved from https://www.r-project.org/about.html
2. Gnu.org. (n.d.). Retrieved 2020, from http://www.gnu.org/philosophy/free-sw.html
3. Li, X. (2013, September). Data Types and Objects in R. Retrieved from https://msu.edu/~lixue/geo866/lab02/data_type.html
4. Contributed Packages. (n.d.). Retrieved from https://cran.r-project.org/web/packages/

Chapter 3: Genetics and Genomics
1. Deoxyribonucleic Acid (DNA). (n.d.). Retrieved from https://www.genome.gov/genetics-glossary/Deoxyribonucleic-Acid
2. What is a gene? - Genetics Home Reference - NIH. (n.d.). Retrieved from https://ghr.nlm.nih.gov/primer/basics/gene
3. Rosalind Franklin: A Crucial Contribution. (n.d.). Retrieved from https://www.nature.com/scitable/topicpage/rosalind-franklin-a-crucial-contribution-6538012/
4. Mitochondrial DNA - Genetics Home Reference - NIH. (n.d.). Retrieved from https://ghr.nlm.nih.gov/mitochondrial-dna
5. Antolin, M., & Black, W. (2013, February 05). Description of Genes. Retrieved from https://www.sciencedirect.com/science/article/pii/B9780123847195000630
6. Cooper GM. The Cell: A Molecular Approach. 2nd edition. Sunderland (MA): Sinauer Associates; 2000. DNA Replication. Available from: https://www.ncbi.nlm.nih.gov/books/NBK9940/

7. Ribonucleic Acid (RNA). (n.d.). Retrieved August, from https://www.genome.gov/genetics-glossary/RNA-Ribonucleic-Acid
8. Transfer RNA / tRNA. (n.d.). Retrieved from https://www.nature.com/scitable/definition/trna-transfer-rna-256/
9. The Editors of Encyclopaedia Britannica. (2020, May 22). Ribosomal RNA. Retrieved August 20, 2020, from https://www.britannica.com/science/ribosomal-RNA
10. Strauss, B. S. (2016). Biochemical Genetics and Molecular Biology: The Contributions of George Beadle and Edward Tatum. *Genetics, 203*(1), 13-20. doi:10.1534/genetics.116.188995
11. Clancy, S. (2008). DNA Transcription. Retrieved from https://www.nature.com/scitable/topicpage/dna-transcription-426/
12. Griffiths AJF, Miller JH, Suzuki DT, et al. An Introduction to Genetic Analysis. 7th edition. New York: W. H. Freeman; 2000. Transcription. Available from: https://www.ncbi.nlm.nih.gov/books/NBK22071/
13. Clancy, S., & Brown, W. (2008). Translation: DNA to mRNA to Protein. Retrieved from https://www.nature.com/scitable/topicpage/translation-dna-to-mrna-to-protein-393/
14. Slonczewski, J. (n.d.). The Genetic Code. Retrieved from http://biology.kenyon.edu/courses/biol114/Chap05/Chapter05.html
15. Rich A, RajBhandary UL. Transfer RNA: molecular structure, sequence, and properties. *Annu Rev Biochem.* 1976;45:805-860. doi:10.1146/annurev.bi.45.070176.004105
16. Lafontaine, D., Tollervey, D. The function and synthesis of ribosomes. *Nat Rev Mol Cell Biol* 2, 514–520 (2001). https://doi.org/10.1038/35080045
17. Pestova, T. V., Kolupaeva, V. G., Lomakin, I. B., Pilipenko, E. V., Shatsky, I. N., Agol, V. I., & Hellen, C. U. (2001). Molecular mechanisms of translation initiation in eukaryotes. *Proceedings of the National*

*Academy of Sciences, 98*(13), 7029-7036. doi:10.1073/pnas.111145798
18. Knight, J. R., Garland, G., Pöyry, T., Mead, E., Vlahov, N., Sfakianos, A., . . . Willis, A. E. (2020). Control of translation elongation in health and disease. *Disease Models & Mechanisms, 13*(3), Dmm043208. doi:10.1242/dmm.043208
19. Berg JM, Tymoczko JL, Stryer L. Biochemistry. 5th edition. New York: W H Freeman; 2002. Chapter 3, Protein Structure and Function. Available from: https://www.ncbi.nlm.nih.gov/books/NBK21177/
20. Phillips, T. (n.d.). Regulation of Transcription and Gene Expression in Eukaryotes. Retrieved from https://www.nature.com/scitable/topicpage/regulation-of-transcription-and-gene-expression-in-1086/
21. Akalin, A. (2020). *Computational Genomics with R.*
22. Filipowicz, W., Bhattacharyya, S. & Sonenberg, N. Mechanisms of post-transcriptional regulation by microRNAs: are the answers in sight?. *Nat Rev Genet* 9, 102–114 (2008). https://doi.org/10.1038/nrg2290
23. Kervestin, S., Amrani, N. Translational regulation of gene expression. *Genome Biol* 5, 359 (2004). https://doi.org/10.1186/gb-2004-5-12-359
24. Lodish H, Berk A, Zipursky SL, et al. Molecular Cell Biology. 4th edition. New York: W. H. Freeman; 2000. Section 8.1, Mutations: Types and Causes. Available from: https://www.ncbi.nlm.nih.gov/books/NBK21578/
25. Griffiths AJF, Miller JH, Suzuki DT, et al. An Introduction to Genetic Analysis. 7th edition. New York: W. H. Freeman; 2000. Aneuploidy. Available from: https://www.ncbi.nlm.nih.gov/books/NBK21870/
26. Stanford Children's Health. (n.d.). Numerical Abnormalities: Overview of Trisomies and Monosomies. Retrieved from https://www.stanfordchildrens.org/en/topic/default?id=numerical-abnormalities-overview-of-trisomies-and-monosomies-90-P02138

27. A Brief Guide to Genomics. (n.d.). Retrieved from https://www.genome.gov/about-genomics/fact-sheets/A-Brief-Guide-to-Genomics
28. Chial, H. (2008). DNA Sequencing Technologies Key to the Human Genome Project. Retrieved from https://www.nature.com/scitable/topicpage/dna-sequencing-technologies-key-to-the-human-828/
29. Collins, F. S. (2001, February 12). Remarks at the Press Conference Announcing Sequencing and Analysis of the Human Genome. Retrieved from https://www.genome.gov/10001379/february-2001-working-draft-of-human-genome-director-collins
30. Venter, J. C., *et al.* The sequence of the human genome. *Science* **291**, 1304–1351 (2001)

Chapter 4: RNA Sequencing and Differential Gene Expression
1. Dundar, F., Skrabanek, L., & Zumbo, P. (2019). *Introduction to differential gene expression analysis using RNA-seq.*
2. Korpelainen, E., Tuimala, J., Somervuo, P., Huss, M., & Wong, G. (2015). *RNA-Seq Data Analysis: A Practical Approach.* Boca Raton: CRC Press, Taylor & Francis Group.
3. Illumina. (n.d.). Simple, customized RNA-Seq workflows. Retrieved from https://www.illumina.com/content/dam/illumina-marketing/documents/products/other/rna-sequencing-workflow-buyers-guide-476-2015-003.pdf
4. Columbia Genome Center. (n.d.). Genome Sequencing: Defining Your Experiment. Retrieved from https://systemsbiology.columbia.edu/genome-sequencing-defining-your-experiment
5. Illumina. (2020, March 10). FASTQ files explained. Retrieved from https://support.illumina.com/bulletins/2016/04/fastq-files-explained.html
6. Dobin, A., & Gingeras, T. R. (2015). Mapping RNA-seq Reads with STAR. *Current protocols in bioinformatics*, *51*, 11.14.1–11.14.19. https://doi.org/10.1002/0471250953.bi1114s51
7. Akalin, A. (2020). *Computational Genomics with R.*

8. MacLean, D. (2019). *R Bioinformatics Cookbook*. Birmingham, UK: Packt Publishing.
9. Michael I. Love, S. (2020). Analyzing RNA-seq data with DESeq2. Retrieved from http://bioconductor.org/packages/devel/bioc/vignettes/DESeq2/inst/doc/DESeq2.html
10. Starmer, J. (2017, March 27). *StatQuest: DESeq2, part 1, Library Normalization* [Video file]. Retrieved from https://www.youtube.com/watch?v=UFB993xufUU
11. Korthauer, K., Kimes, P.K., Duvallet, C. et al. A practical guide to methods controlling false discoveries in computational biology. *Genome Biol* 20, 118 (2019). https://doi.org/10.1186/s13059-019-1716-1
12. Robinson, M. D., McCarthy, D. J., & Smyth, G. K. (2010). edgeR: a Bioconductor package for differential expression analysis of digital gene expression data. *Bioinformatics (Oxford, England)*, *26*(1), 139–140. https://doi.org/10.1093/bioinformatics/btp616
13. Starmer, J. (2017, April 3). *StatQuest: EdgeR, part 1, Library Normalization* [Video file]. Retrieved from https://www.youtube.com/watch?v=Wdt6jdi-NQo
14. Lever, J., Krzywinski, M. & Altman, N. Principal component analysis. *Nat Methods* 14, 641–642 (2017). https://doi.org/10.1038/nmeth.4346
15. Piech, C. (2013). K Means. Retrieved from https://stanford.edu/~cpiech/cs221/handouts/kmeans.html
16. Kim, S., Park, Y., & Teng, G. C. (2015). MetaOmics: Transcriptomic meta-analysis methods for biomarker detection, pathway analysis and other exploratory purposes. In Z. Huo (Ed.), *Integrating Omics Data*. Cambridge University Press.

Made in the USA
Las Vegas, NV
03 February 2021

17136578R00071